MATHEMATICAL STRUCTURES
of
QUANTUM MECHANICS

MATHEMATICAL STRUCTURES
QUANTUM MECHANICS

Kow Lung CHANG
National Taiwan University, Taiwan

World Scientific

NEW JERSEY · LONDON · SINGAPORE · BEIJING · SHANGHAI · HONG KONG · TAIPEI · CHENNAI

Published by

World Scientific Publishing Co. Pte. Ltd.

5 Toh Tuck Link, Singapore 596224

USA office: 27 Warren Street, Suite 401-402, Hackensack, NJ 07601

UK office: 57 Shelton Street, Covent Garden, London WC2H 9HE

British Library Cataloguing-in-Publication Data

A catalogue record for this book is available from the British Library.

MATHEMATICAL STRUCTURES OF QUANTUM MECHANICS

ISBN-13 978-981-4366-58-8

ISBN-10 981-4366-58-7

Printed in Singapore by World Scientific Printers.

TO FEZA AND SUHA

Preface

During the past few years, after a couple of weeks of lecturing the course of quantum mechanics that I offered at the Physics Department, National Taiwan University, some students would usually come to ask me as to what extent they had to refurbish their mathematical background in order to follow my lecture with ease and confidence. It was hard for me to provide a decent and proper answer to the question, and very often students would show reluctance to invest extra time on subjects such as group theory or functional analysis when I advised them to take some advanced mathematics courses. All these experiences that I have encountered in my class eventually motivated me to write this book.

The book is designed with the hope that it might be helpful to those students I mentioned above. It could also serve as a complementary text in quantum mechanics for students of inquiring minds who appreciate the rigor and beauty of quantum theory.

Assistance received from many sources made the appearance of this book possible. I wish to express here my great appreciation and gratitude to Dr. Yusuf Gürsey, who painstakingly went through the manuscript and responded generously by giving very helpful suggestions and comments, and made corrections line by line. I would also like to thank Mr. Paul Black who provided me with cogent suggestions and criticism of the manuscript, particularly in those sections on quantum uncertainty. I am indebted as well to Mr. Chih Han Lin who, with immense patience, compiled the whole text and drew all the figures from my suggestions. All his hard work and attention resulted in the present form of this book.

Taipei, Taiwan
March, 2011   Kow Lung Chang

Contents

Chapter 1

Postulates and Principles of Quantum Mechanics

As with many fields in physics, a precise and rigorous description of a given subject requires the use of some mathematical tools. Take Lagrange's formulation of classical mechanics for instance, one needs the basic knowledge of variational calculus in order to derive the equations of motion for a system of particles in terms of generalized coordinates. To formulate the postulates of quantum mechanics, it would also be necessary to acquire some knowledge on vector space in general, and Hilbert space in particular. It is in this chapter that we shall provide the minimum but essential mathematical preparation that allows one to perceive and understand the general framework of quantum theory and to appreciate the rigorous derivation of the quantum principles.

1.1 Vector space

A **vector space** $\mathcal{V}$ is a set of elements, called vectors, with the following 2 operations:

- An operation of addition, which for each pair of vectors ψ and ϕ, corresponds to a new vector $\psi + \phi \in \mathcal{V}$, called the sum of ψ and ϕ.

- An operation of scalar multiplication, which for each vector ψ and a number a, specifies a vector $a\psi$, such that (assuming a, b are numbers and ψ, ϕ and χ are vectors)

1

$$\psi + \phi = \phi + \psi, \tag{1.1a}$$

$$\psi + (\phi + \chi) = (\psi + \phi) + \chi, \tag{1.1b}$$

$$\psi + 0 = \psi, \quad 0 \text{ is null vector}, \tag{1.1c}$$

$$a(\psi + \phi) = a\psi + a\phi, \tag{1.1d}$$

$$(a + b)\psi = a\psi + b\psi, \tag{1.1e}$$

$$a(b\psi) = (ab)\psi, \tag{1.1f}$$

$$1 \cdot \psi = \psi, \tag{1.1g}$$

$$0 \cdot \psi = 0, \tag{1.1h}$$

where if a, b are real numbers, we call this vector space the real vector space, and denote it by $\mathcal{V}_r$. On the other way, complex vector space $\mathcal{V}_c$ means a, b are complex numbers.

Example

We take n-dimensional Euclidean space, $\mathcal{R}^n$-space, as an exmple. It is a vector space with the vectors ψ and ϕ specified as $\psi = (x_1, x_2, \ldots, x_i, \ldots, x_n)$ and $\phi = (y_1, y_2, \ldots, y_i, \ldots, y_n)$, where x_i and y_i ($i = 1, 2, \ldots, n$) are all taken as real numbers. The sum of ψ and ϕ becomes $(x_1 + y_1, x_2+y_2, \ldots, x_i+y_i, \ldots, x_n+y_n)$ and $a\psi = (ax_1, ax_2, \ldots, ax_i, \ldots, ax_n)$. If a and x_i are taken as complex numbers, then ψ is a vector in $\mathcal{C}^n$-space; a n-dimensional complex vector space.

It is easily understood that a set of the continuous functions $f(x)$ for $a \leqslant x \leqslant b$ forms a vector space, namely $\mathcal{L}^2(a, b)$-space.

Before leaving this section, we also introduce some terminologies in the following subsections that will be frequently referred to in later chapters.

1.1.1 Linearly dependent and linearly independent

Consider a set of m vectors $\{\psi_1, \psi_2, \ldots, \psi_m\}$, and we construct the linear combination of these m vectors as follows:

$$\sum_{i=1}^{m} a_i \psi_i. \tag{1.2}$$

This linear combination of m vectors is of course a vector. It becomes a **null vector** if and only if all the coefficient $a_i = 0$ for $i = 1, 2, \ldots, m$, then the set of m vectors $\{\psi_1, \psi_2, \ldots, \psi_m\}$ is called linearly independent. If at least one of the coefficient $a_l \neq 0$ such that $\sum_{i=1}^{m} a_i \psi_i = 0$, then the set $\{\psi_1, \psi_2, \ldots, \psi_m\}$ is called linearly dependent.

1.1.2 Dimension and basis

The maximum number of linearly independent vectors in $\mathcal{V}$ is called the **dimension** of $\mathcal{V}$. Any n-linearly independent vectors in vector space $\mathcal{V}$ of n-dimension form the **basis** of the vector space.

1.2 Inner product

An **inner product**, or sometimes called **scalar product** in vector space, is a numerically valued function of the ordered pair of vectors ψ and ϕ, denoted by (ψ, ϕ), and for a scalar a, such that

$$(\psi, \phi + \chi) = (\psi, \phi) + (\psi, \chi), \tag{1.3a}$$
$$(\psi, a\phi) = a(\psi, \phi), \tag{1.3b}$$
$$(\psi, \phi) = (\phi, \psi)^*, \tag{1.3c}$$
$$(\psi, \psi) \geqslant 0, (\psi, \psi) = 0 \text{ if and only if } \psi \text{ is a null vector.} \tag{1.3d}$$

Two vectors ψ and ϕ are said to be orthogonal to each other if their corresponding inner product vanishes, namely $(\psi, \phi) = 0$.

For example, let us consider the vectors in $\mathcal{C}^n$-space $\psi = (x_1, x_2, \ldots, x_n)$ and $\phi = (y_1, y_2, \ldots, y_n)$ where x_i and y_i are complex numbers. The inner product of ψ and ϕ written as

$$(\psi, \phi) = \sum_{i=1}^{n} x_i^* y_i = x_1^* y_1 + x_2^* y_2 + \cdots + x_n^* y_n. \tag{1.4}$$

Consider the set of continuous function of $f(x)$ where $a \leqslant x \leqslant b$. An ordered pair of functions $f(x)$ and $g(x)$ define the inner product as $(f(x), g(x)) = \int_a^b f(x)^* g(x) dx$. This vector space is called $\mathcal{L}^2(a, b)$-space when $|f(x)|^2$ and $|g(x)|^2 = $ finite.

1.2.1 Schwarz inequality

We are now in the position to prove the Schwarz inequality.

Let ψ and ϕ be any two vectors. The **Schwarz inequality** reads as

$$|(\psi,\phi)| = \sqrt{(\psi,\phi)(\phi,\psi)} \leqslant \sqrt{(\psi,\psi)}\sqrt{(\phi,\phi)}. \qquad (1.5)$$

Proof

Since $(\psi + \alpha\phi, \psi + \alpha\phi) \geqslant 0$, where $\alpha = \xi + i\eta$ is a complex number. Regard this inner product $(\psi + \alpha\phi, \psi + \alpha\phi) = f(\xi,\eta)$ as a function of two variables ξ and η. Then

$$f(\xi,\eta) = (\psi,\psi) + |\alpha|^2(\phi,\phi) + \alpha(\psi,\phi) + \alpha^*(\phi,\psi), \qquad (1.6)$$

which is positive definite. Let us look for the minimum of $f(\xi,\eta)$ at ξ_0, η_0 by solving

$$\left.\frac{\partial f(\xi,\eta)}{\partial \xi}\right|_{\xi_0,\eta_0} = \left.\frac{\partial f(\xi,\eta)}{\partial \eta}\right|_{\xi_0,\eta_0} = 0, \qquad (1.7)$$

and we obtain

$$\xi_0 = \frac{1}{2}\frac{(\psi,\phi) + (\phi,\psi)}{(\phi,\phi)}, \quad \eta_0 = -\frac{i}{2}\frac{(\psi,\phi) - (\phi,\psi)}{(\phi,\phi)}. \qquad (1.8)$$

Therefore

$$f(\xi_0,\eta_0) = (\psi,\psi) - \frac{(\psi,\phi)(\phi,\psi)}{(\phi,\phi)} \geqslant 0, \qquad (1.9)$$

that can be cast into the familiar expression of Schwarz inequality.

1.2.2 Gram-Schmidt orthogonalization process

The inner product we have been considering can be applied to the orthogonalization of the basis in the n-dimensional vector space. Let

$\{\psi_1, \psi_2, \ldots, \psi_n\} \in \mathcal{V}$ be the set of n-linearly independent vectors. Since $(\psi_i, \psi_j) \neq 0$ in general, we can construct a new set of vectors $\{\psi_1', \psi_2', \ldots, \psi_n'\}$ such that $(\psi_i', \psi_j') = 0$ for all i and j unless $i = j$, namely ψ_i' and ψ_j' are orthogonal to each other for $i \neq j$ by the following procedure:

First take $\psi_1' = \psi_1$ and construct $\psi_2' = \psi_2 + \alpha\psi_1'$. In order to force ψ_2' to be orthogonal to ψ_1', we solve the α as to meet the condition $(\psi_2', \psi_1') = 0$, i.e.

$$(\psi_2', \psi_1') = (\psi_2, \psi_1') + \alpha^*(\psi_1', \psi_1') = 0, \tag{1.10}$$

and we obtain $\alpha = -(\psi_2, \psi_1')^*/(\psi_1', \psi_1') = -(\psi_1', \psi_2)/(\psi_1', \psi_1')$, hence

$$\psi_2' = \psi_2 - \psi_1'\frac{(\psi_1', \psi_2)}{(\psi_1', \psi_1')}. \tag{1.11}$$

The same procedure can be performed repeatedly to reach $\psi_3' = \psi_3 + \alpha\psi_2' + \beta\psi_1'$ which guarantees $(\psi_3', \psi_1') = (\psi_3', \psi_2') = 0$ with $\alpha = -(\psi_2', \psi_3)/(\psi_2', \psi_2')$ and $\beta = -(\psi_1', \psi_3)/(\psi_1', \psi_1')$. In general,

$$\psi_i' = \psi_i - \psi_{i-1}'\frac{(\psi_{i-1}', \psi_i)}{(\psi_{i-1}', \psi_{i-1}')} - \psi_{i-2}'\frac{(\psi_{i-2}', \psi_i)}{(\psi_{i-2}', \psi_{i-2}')} - \cdots - \psi_1'\frac{(\psi_1', \psi_i)}{(\psi_1', \psi_1')}.$$

$$\tag{1.12}$$

The set of orthogonal basis $\{\psi_1', \psi_2', \ldots, \psi_n'\}$ can be normalized immediately by multiplying the inverse square root of the corresponding inner product, i.e.

$$\tilde{\psi}_i = \frac{\psi_i'}{\sqrt{(\psi_i', \psi_i')}}, \tag{1.13}$$

and $\{\tilde{\psi}_1, \tilde{\psi}_2, \ldots, \tilde{\psi}_n\}$ becomes the **orthonormal** set of the basis in the vector space. From now on we shall take the basis to be orthonormal without mentioning it particularly.

Example

Consider the following set of continuous functions in $C(-\infty, \infty)$

$$f_n(x) = x^n \exp\left(-\frac{x^2}{2}\right), \quad n = 0, 1, \ldots . \tag{1.14}$$

We construct the new set of orthogonal vectors by applying the Gram-Schmidt process and obtain:

$$f_0'(x) = f_0(x) = \exp\left(-\frac{x}{2}\right), \tag{1.15}$$

$$f_1'(x) = f_1 - \frac{f_0'(f_0', f_1)}{(f_0', f_0')} = f_1(x) = x \exp\left(\frac{x^2}{2}\right), \tag{1.16}$$

$$f_2'(x) = f_2 - \frac{f_1'(f_1', f_2)}{(f_1', f_1')} - \frac{f_0'(f_0', f_2)}{(f_0', f_0')} = \left(x^2 - \frac{1}{2}\right) \exp\left(-\frac{x^2}{2}\right). \tag{1.17}$$

Similarly we have $f_3'(x) = (x^3 - 3x/2) \exp(-x^2/2)$. The orthonormal functions can be calculated according to

$$\tilde{f}_n(x) = \frac{f_n'(x)}{\sqrt{(f_n'(x), f_n'(x))}} = \frac{1}{\sqrt{2^n n! \sqrt{\pi}}} \exp\left(-\frac{x^2}{2}\right) H_n(x), \tag{1.18}$$

where $H_n(x)$ are called Hermite polynomials. One also recognizes that $\tilde{f}_n(x)$ are in fact, the eigenfunctions of the Schrödinger equation for one-dimension harmonic oscillation.

1.3 Completeness and Hilbert space

Let us introduce some other terminologies in discussing Hilbert space.

1.3.1 Norm

A **norm** on a vector space is a non-negative real function such that, if ψ, ϕ are vectors, the norm of ψ is written as $\|\psi\|$, satisfying:

$$\|\psi\| \geqslant 0, \qquad \|\psi\| = 0 \quad \text{iff} \quad \psi \text{ is null vector,} \qquad (1.19a)$$

$$\|a\psi\| = |a| \cdot \|\psi\|, \qquad (1.19b)$$

$$\|\psi + \phi\| \leqslant \|\psi\| + \|\phi\|. \qquad (1.19c)$$

Example

If $f(x) \in C(a, b)$, namely if $f(x)$ is a continuous function for a variable that lies between a and b, the norm of $f(x)$ can be defined either as $\|f(x)\| = \text{Max}\{|f(x)|, a \leqslant x \leqslant b\}$ or as the inner product of $f(x)$, i.e.

$$\|f(x)\|^2 = (f(x), f(x)) = \int_a^b |f(x)|^2 dx.$$

1.3.2 Cauchy sequence and convergent sequence

Consider an infinite dimensional vector space and denote the basis by $\{\phi_1, \phi_2, \phi_3, \ldots\}$. We construct the partial sum $\psi_N = \sum_i a_i \phi_i$, where i runs from 1 to N, and obtain $\ldots, \psi_j, \psi_{j+1}, \ldots, \psi_m, \psi_{m+1}, \ldots, \psi_n, \ldots$ for increasing values in N that forms an infinite sequence. The sequence is called a **Cauchy sequence** if

$$\lim_{n \to \infty} \psi_n = \lim_{m \to \infty} \psi_m,$$

or more precisely to put in terms of norm, i.e, $\lim\limits_{n,m \to \infty} \|\psi_n - \psi_m\| = 0$.

It is said that a vector ψ_m converges to ψ if

$$\lim_{m \to \infty} \psi_m = \psi, \quad \text{or} \quad \lim_{m \to \infty} \|\psi_m - \psi\| = 0,$$

then $\{\ldots, \psi_{m-1}, \psi_m, \ldots\}$ is called a **convergent sequence**.

It is easily concluded that every convergent sequence is a Cauchy sequence. Yet it is not necessary true conversely. Namely a Cauchy sequence is not always a convergent sequence.

1.3.3 Complete vector space

A vector space, in which every Cauchy sequence of a vector ψ_m converges to a limiting vector ψ, is called a complete vector space.

1.3.4 Hilbert space

A Hilbert space is a complete vector space with norm defined as the inner product. A Hilbert space, finite dimensional or infinite dimensional, is separable if its basis is countable.

1.4 Linear operator

A **linear operator A** on a vector space assigns to each vector ψ a new vector, i.e. $\mathbf{A}\psi = \psi'$ such that

$$\mathbf{A}(\psi + \phi) = \mathbf{A}\psi + \mathbf{A}\phi, \quad \mathbf{A}(\alpha\psi) = \alpha\mathbf{A}\psi. \tag{1.20}$$

Two operators $\mathbf{A}$, $\mathbf{B}$ are said equal if $\mathbf{A}\psi = \mathbf{B}\psi$ for all ψ in the vector space.

For convenience in later discussion, we denote

- **O**: null operator such that $\mathbf{O}\psi = 0$ for all ψ, and 0 is the null vector.

- **I**: unit operator or identity operator such that $\mathbf{I}\psi = \psi$.

The sum of the operators $\mathbf{A}$ and $\mathbf{B}$ is an operator, such that $(\mathbf{A}+\mathbf{B})\psi = \mathbf{A}\psi + \mathbf{B}\psi$. The product of operators $\mathbf{A}$ and $\mathbf{B}$ is again an operator that one writes as $\mathbf{A} \cdot \mathbf{B}$ or $\mathbf{AB}$ such that $(\mathbf{AB})\psi = \mathbf{A}(\mathbf{B}\psi)$.

The order of the operators in the product matters greatly. It is generally that $\mathbf{AB} \neq \mathbf{BA}$. The associative rule holds for the product of the operators $\mathbf{A}(\mathbf{BC}) = (\mathbf{AB})\mathbf{C}$.

1.4.1 Bounded operator

An operator $\mathbf{A}$ is called a **bounded operator** if there exists a positive number b such that

$$\|\mathbf{A}\psi\| \leqslant b\|\psi\|, \quad \text{for any vector } \psi \text{ in the vector space.}$$

The least upperbound (supremum) of $\mathbf{A}$, namely the smalleast number of b for a given operator $\mathbf{A}$ and for any ψ in $\mathcal{V}$, is denoted by

$$\|\mathbf{A}\| = \sup\left\{\frac{\|\mathbf{A}\psi\|}{\|\psi\|}, \quad \psi \neq 0\right\}, \tag{1.21}$$

then $\|\mathbf{A}\psi\| \leqslant \|\mathbf{A}\|\|\psi\|$.

We are now able to show readily that $\|\mathbf{A} + \mathbf{B}\| \leqslant \|\mathbf{A}\| + \|\mathbf{B}\|$.

Proof

Let us denote $\|\mathbf{A}\psi\| \leqslant \|\mathbf{A}\|\|\psi\|$ and $\|\mathbf{B}\psi\| \leqslant \|\mathbf{B}\|\|\psi\|$. Then

$$\|\mathbf{A} + \mathbf{B}\| = \sup\left\{\frac{\|(\mathbf{A}+\mathbf{B})\psi\|}{\|\psi\|}, \psi \neq 0\right\} = \sup\left\{\frac{\|\mathbf{A}\psi + \mathbf{B}\psi\|}{\|\psi\|}, \psi \neq 0\right\}$$

$$\leqslant \sup\left\{\frac{\|\mathbf{A}\psi\|}{\|\psi\|}, \psi \neq 0\right\} + \sup\left\{\frac{\|\mathbf{B}\psi\|}{\|\psi\|}, \psi \neq 0\right\} = \|\mathbf{A}\| + \|\mathbf{B}\|.$$

Similarly, we have $\|\mathbf{AB}\| \leqslant \|\mathbf{A}\|\|\mathbf{B}\|$.

1.4.2 Continuous operator

Consider the convergent sequence $\{\ldots, \psi_m, \psi_{m+1}, \ldots, \psi_n, \ldots\}$ such that $\lim\limits_{n\to\infty} \|\psi_n - \psi\| = 0$. If $\mathbf{A}$ is a bounded operator, then $\{\ldots, \mathbf{A}\psi_m,$ $\mathbf{A}\psi_{m+1}, \ldots, \mathbf{A}\psi_n, \ldots\}$ is also a convergent sequence because

$$\lim_{n\to\infty} \|\mathbf{A}\psi_n - \mathbf{A}\psi\| \leqslant \|\mathbf{A}\| \lim_{n\to\infty} \|\psi_n - \psi\| = 0.$$

We call operator $\mathbf{A}$ the **continuous operator**.

1.4.3 Inverse operator

An operator $\mathbf{A}$ has an **inverse operator** if there exists $\mathbf{B}_R$ such that $\mathbf{A}\mathbf{B}_R = \mathbf{I}$, then we call operator $\mathbf{B}_R$ the right inverse of $\mathbf{A}$. Similarly an operator $\mathbf{B}_L$ such that the product operator $\mathbf{B}_L\mathbf{A} = \mathbf{I}$, then we call operator $\mathbf{B}_L$ the left inverse of $\mathbf{A}$. In fact, the left inverse operator is always equal to the right inverse operator for a given operator $\mathbf{A}$, because

$$\mathbf{B}_L = \mathbf{B}_L\mathbf{I} = \mathbf{B}_L(\mathbf{A}\mathbf{B}_R) = (\mathbf{B}_L\mathbf{A})\mathbf{B}_R = \mathbf{I}\mathbf{B}_R = \mathbf{B}_R. \tag{1.22}$$

The inverse operator of a given operator $\mathbf{A}$ is also unique. If operators $\mathbf{B}$ and $\mathbf{C}$ are all inverse operators of $\mathbf{A}$, then $\mathbf{C} = \mathbf{C}\mathbf{I} = \mathbf{C}(\mathbf{A}\mathbf{B}) = (\mathbf{C}\mathbf{A})\mathbf{B} = \mathbf{B}$.

The implication of uniqueness of the inverse operator of operator $\mathbf{A}$ allows us to write it in the form $\mathbf{A}^{-1}$, namely $\mathbf{A}\mathbf{A}^{-1} = \mathbf{A}^{-1}\mathbf{A} = \mathbf{I}$. It is easily verified that $(\mathbf{A}\mathbf{B})^{-1} = \mathbf{B}^{-1}\mathbf{A}^{-1}$.

1.4.4 Unitary operator

An operator $\mathbf{U}$ is unitary if $\|\mathbf{U}\psi\| = \|\psi\|$. A unitary operation preserves the invariant of the inner product of any pair of vectors, i.e. $(\mathbf{U}\psi, \mathbf{U}\phi) = (\psi, \phi)$. This can be proved as follows:

Let $\chi = \psi + \phi$ and we have

$$\begin{aligned}
(\mathbf{U}\chi, \mathbf{U}\chi) &= (\mathbf{U}(\psi + \phi), \mathbf{U}(\psi + \phi)) \\
&= (\mathbf{U}\psi, \mathbf{U}\psi) + (\mathbf{U}\psi, \mathbf{U}\phi) + (\mathbf{U}\phi, \mathbf{U}\psi) + (\mathbf{U}\phi, \mathbf{U}\phi) \\
&= \|\mathbf{U}\psi\|^2 + \|\mathbf{U}\phi\|^2 + 2\Re\{(\mathbf{U}\psi, \mathbf{U}\phi)\},
\end{aligned}$$

and on the other hand,

$$\begin{aligned}
(\mathbf{U}\chi, \mathbf{U}\chi) &= (\psi + \phi, \psi + \phi) \\
&= (\chi, \chi) = (\psi, \psi) + (\psi, \phi) + (\phi, \psi) + (\phi, \phi) \\
&= \|\psi\|^2 + \|\phi\|^2 + 2\Re\{(\psi, \phi)\}.
\end{aligned}$$

Since $\|\mathbf{U}\psi\| = \|\psi\|, \|\mathbf{U}\phi\| = \|\phi\|$, we have $\Re\{(\mathbf{U}\psi, \mathbf{U}\phi)\} = \Re\{(\psi, \phi)\}$. Similarly if $\chi' = \psi + i\phi$, we obtain $(\mathbf{U}\chi', \mathbf{U}\chi') = (\chi', \chi')$, that implies $\Im\{(\mathbf{U}\psi, \mathbf{U}\phi)\} = \Im\{(\psi, \phi)\}$, therefore $(\mathbf{U}\psi, \mathbf{U}\phi) = (\psi, \phi)$.

1.4.5 Adjoint operator

Consider the inner product of $(\psi, \mathbf{A}\phi)$ where $\mathbf{A}$ is a given linear operator of interest. This numerically scalar quantity certainly is a function of operator $\mathbf{A}$ and the pair of vectors ψ and ϕ, namely $(\psi, \mathbf{A}\phi) = F(\mathbf{A}, \psi, \phi)$ is a scalar quantity.

Instead of performing the above inner product straightforwardly, we shall obtain the very same scalar of $(\psi, \mathbf{A}\phi)$ by forming the following inner product $(\mathbf{A}^{\dagger}\psi, \phi)$ such that $(\psi, \mathbf{A}\phi) \equiv (\mathbf{A}^{\dagger}\psi, \phi)$. The operator $\mathbf{A}^{\dagger}$ is called the **adjoint operator** of $\mathbf{A}$. The following relations can be easily established (proofs left to readers):

$$(\mathbf{A} + \mathbf{B})^{\dagger} = \mathbf{A}^{\dagger} + \mathbf{B}^{\dagger}, \tag{1.23a}$$

$$(\alpha\mathbf{A})^{\dagger} = \alpha^{*}\mathbf{A}^{\dagger}, \tag{1.23b}$$

$$(\mathbf{A}\mathbf{B})^{\dagger} = \mathbf{B}^{\dagger}\mathbf{A}^{\dagger}, \tag{1.23c}$$

$$(\mathbf{A}^{\dagger})^{\dagger} = \mathbf{A}, \tag{1.23d}$$

$$(\mathbf{A}^{\dagger})^{-1} = (\mathbf{A}^{-1})^{\dagger}. \tag{1.23e}$$

It can also be shown that $\mathbf{A}^{\dagger}$ is a bounded operator if $\mathbf{A}$ is bounded and their norms are equal, i.e. $\|A\| = \|A^{\dagger}\|$.

To prove the above equality, let us consider $\|\mathbf{A}^{\dagger}\psi\|^{2} = (\mathbf{A}^{\dagger}\psi, \mathbf{A}^{\dagger}\psi)$, namely

$$\|\mathbf{A}^{\dagger}\psi\|^{2} = (\mathbf{A}^{\dagger}\psi, \mathbf{A}^{\dagger}\psi) = (\mathbf{A}\mathbf{A}^{\dagger}\psi, \psi) \leqslant \|\psi\|\|\mathbf{A}\mathbf{A}^{\dagger}\psi\| \leqslant \|\psi\|\|\mathbf{A}\|\|\mathbf{A}^{\dagger}\psi\|,$$

therefore $\|\mathbf{A}^{\dagger}\psi\| \leqslant \|\mathbf{A}\|\|\psi\|$, and we have $\|\mathbf{A}^{\dagger}\| \leqslant \|\mathbf{A}\|$.

On the other hand, we have $\|\mathbf{A}\psi\|^{2} = (\mathbf{A}\psi, \mathbf{A}\psi) = (\mathbf{A}^{\dagger}\mathbf{A}\psi, \psi) \leqslant \|\psi\|\|\mathbf{A}^{\dagger}\|\|\mathbf{A}\psi\|$, which implies $\|\mathbf{A}\| \leqslant \|\mathbf{A}^{\dagger}\|$. Therefore $\|A\| = \|A^{\dagger}\|$ is established.

1.4.6 Hermitian operator

When an operator is self-adjoint, namely an adjoint operator $\mathbf{A}^\dagger$ equals to operator $\mathbf{A}$ itself, i.e. $\mathbf{A} = \mathbf{A}^\dagger$, then we call $\mathbf{A}$ a **Hermitian operator**.

1.4.7 Projection operator

Let $\mathcal{H}$ be a Hilbert space in which we consider a **subspace** $\mathcal{M}$ and its orthogonal complement space $\mathcal{M}_\perp$ such that for each vector ψ in $\mathcal{H} = \mathcal{M} \oplus \mathcal{M}_\perp$ that are decomposed into unique vectors $\psi_\mathcal{M}$ in $\mathcal{M}$ and $\psi_{\mathcal{M}_\perp}$ in $\mathcal{M}_\perp$ such that $\psi = \psi_\mathcal{M} + \psi_{\mathcal{M}_\perp}$, and $(\psi_\mathcal{M}, \psi_{\mathcal{M}_\perp}) = 0$.

The projection operator $\mathbf{P}_\mathcal{M}$ when acting upon vector ψ onto a subspace results in $\mathbf{P}_\mathcal{M}\psi = \psi_\mathcal{M}$. It is obvious that $\mathbf{P}_\mathcal{M}\psi = \psi$ if $\psi \in \mathcal{M}$ and $\mathbf{P}_\mathcal{M}\psi = 0$ if $\psi \in \mathcal{M}_\perp$.

One can also be easily convinced that

$$(\psi, \mathbf{P}_\mathcal{M}\phi) = (\psi, \phi_\mathcal{M}) = (\psi_\mathcal{M} + \psi_{\mathcal{M}\perp}, \phi_\mathcal{M}) = (\psi_\mathcal{M}, \phi_\mathcal{M})$$
$$= (\psi_\mathcal{M}, \phi) = (\mathbf{P}_\mathcal{M}\psi_\mathcal{M}, \phi) = (\mathbf{P}_\mathcal{M}\psi, \phi).$$

Therefore $\mathbf{P}_\mathcal{M}$ is also a Hermitian operator, i.e. $\mathbf{P}^\dagger_\mathcal{M} = \mathbf{P}_\mathcal{M}$.

Similarly we define $\mathbf{P}_{\mathcal{M}_\perp}$ such that $\mathbf{P}_{\mathcal{M}_\perp}\psi = \psi_{\mathcal{M}_\perp}$ and the sum of $\mathbf{P}_\mathcal{M}$ and $\mathbf{P}_{\mathcal{M}_\perp}$ becomes an identity operator, i.e.

$$\mathbf{P}_\mathcal{M} + \mathbf{P}_{\mathcal{M}_\perp} = \mathbf{I}.$$

1.4.8 Idempotent operator

The **projection operator** is an **idempotent operator**, namely $\mathbf{P}^2_\mathcal{M} = \mathbf{P}_\mathcal{M}$ because $\mathbf{P}^2_\mathcal{M}\psi = \mathbf{P}_\mathcal{M}\psi_\mathcal{M} = \mathbf{P}_\mathcal{M}\psi$.

1.5 The postulates of quantum mechanics

We start to formulate the postulates of quantum mechanics. We shall treat the first three postulates in this chapter, and leave the 4th postulate for the next chapter when we investigate the time evolution of a quantum system.

1*st* postulate of quantum mechanics:

For every physical system, there exists an abstract entity, called the state (or the state function or wave function that shall be discussed later), which provides the information of the dynamical quantities of the system; such as coordinates, momenta, energy, angular momentum, charge or isospin, etc. All the states for a given physical system are elements of a Hilbert space, i.e.

physical system	$\longleftrightarrow$	Hilbert space $\mathcal{H}$
physical state	$\longleftrightarrow$	state vector ψ in $\mathcal{H}$

Furthermore for each physical observable, such as the 3rd component of the angular momentum or the total energy of the system and so forth, there associates a unique Hermitian operator in the Hilbert space, i.e.

physical (dynamical) observable		corresponding hermitean operator
total energy E	$\longleftrightarrow$	$\mathbf{H} = \mathbf{H}^\dagger$
coordinate $\vec{x}$	$\longleftrightarrow$	$\mathbf{X} = \mathbf{X}^\dagger$
angular momentum $\vec{l}$	$\longleftrightarrow$	$\mathbf{L} = \mathbf{L}^\dagger$

The physical quantity measured in the system for the corresponding observable is obtained by taking the inner product of the pair ψ and $\mathbf{A}\psi$, i.e.

$$\langle \mathbf{A} \rangle = (\psi, \mathbf{A}\psi), \qquad (1.24)$$

which is called the **expectation value** of dynamical quantity $\mathbf{A}$ for the system in the state ψ, which is normalized, i.e. $\|\psi\| = 1$.

Since the action of operator $\mathbf{A}$ upon the vector ψ changes it into another vector ϕ, which implies that the action of the measurement of the dynamical quantity in a certain state usually would disturb the physical system and the original state is changed into another state due to the external disturbance accompanying the measurement.

In particular, if an operator $\mathbf{A}$ such that $\mathbf{A}\psi_a = a\psi_a$, i.e. when $\mathbf{A}$ acts upon a particular physical state ψ_a, the resultant state is the same as the one before, then it is said that the physical state is prepared for the measurement of the **dynamical observable** associated with the operator $\mathbf{A}$. We shall name:

- ψ_a : the state particularly prepared in the system for the measurement of the dynamic quantity, called the **eigenstate** of the operator $\mathbf{A}$.

- a : the value of the measurement of the dynamical quantity in the particular prepared state, called the **eigenvalue** of the operator $\mathbf{A}$.

We shall now explore some properties concerning the eigenvectors and the eigenvalues through a few propositions.

Proposition 1.

The eigenvalues for a Hermitian operator are all real.

Let $\mathbf{A}\psi_a = a\psi_a$ and $\mathbf{A}^\dagger\psi_a = a\psi_a$, and consider the inner product $\langle\mathbf{A}\rangle_{\psi_a} = (\psi_a, \mathbf{A}\psi_a) = (\psi_a, a\psi_a) = a(\psi_a, \psi_a) = a$. On the other hand, we have $\langle\mathbf{A}\rangle_{\psi_a} = (\mathbf{A}^\dagger\psi_a, \psi_a) = (\mathbf{A}\psi_a, \psi_a) = a^*(\psi_a, \psi_a) = a^*$ which implies $a = a^*$ if ψ_a is not a null vector.

Proposition 2.

Two eigenvectors of a Hermitian operator are orthogonal to each other if the corresponding eigenvalues are unequal.

Let $\mathbf{A}\psi_a = a\psi_a$ and $\mathbf{A}\psi_b = b\psi_b$ where $\psi_a \neq \psi_b$, and since

$$(\psi_a, \mathbf{A}\psi_b) = b(\psi_a, \psi_b) = (\mathbf{A}^\dagger\psi_a, \psi_b) = a^*(\psi_a, \psi_b) = a(\psi_a, \psi_b),$$

therefore $(a - b)(\psi_a, \psi_b) = 0$. That implies $(\psi_a, \psi_b) = 0$ if $a \neq b$.

It often occurs that there exists more than one eigenvector of an operator with the same eigenvalue. Consider the Hermitian operator $\mathbf{C}$, such that

$$\mathbf{C}\psi_{c_i} = c_i\psi_{c_i}, \tag{1.25}$$

where $c_1 = c_2 = \ldots = c_m = c$, and $(\psi_{c_1}, \psi_{c_2}, \ldots, \psi_{c_m})$ are linearly independent.

The eigenvalue c is called m-fold degenerate if there are m linearly independent eigenvectors corresponding to the same eigenvalue c of the operator $\mathbf{C}$.

Proposition 3.

If the eigenvalue c of the operator $\mathbf{C}$ is degenerate, any linear combination of the linearly independent eigenvectors is also an eigenvector.

Due to the linearity of operator $\mathbf{C}$, the linear combination $\sum \alpha_i \psi_{c_i}$ is also the eigenvector of $\mathbf{C}$ with the eigenvalue c. By means of the Gram-Schmidt orthogonalization process, one is able to easily construct a new set of orthonormal vectors $\{\tilde{\psi}_{c_1}, \tilde{\psi}_{c_2}, \ldots, \tilde{\psi}_{c_m}\}$ out of the previous m linearly independent set $\{\psi_{c_1}, \psi_{c_2}, \ldots, \psi_{c_m}\}$, such that

$$(\tilde{\psi}_{c_i}, \tilde{\psi}_{c_j},) = \delta_{ij}. \tag{1.26}$$

Following the results of Propositions 2 and 3, we conclude that

$$(\tilde{\psi}_{a_i}, \tilde{\psi}_{b_j},) = \delta_{ab}\delta_{ij}, \tag{1.27}$$

where a, b refer to the eigenvalues and i, j refer to the index of degeneracy. We shall drop $\tilde{}$ on top of the orthonormal basis without mentioning it further.

2nd postulate of quantum mechanics:

The set of eigenvectors ψ_{a_i} of a given Hermitian operator corresponding to a physical observable form the basis of a Hilbert space. Any state in the physical system can be denoted by a vector in the Hilbert space as a linear combination of ψ_{a_i}, i.e. $\psi = \sum \alpha_{a_i} \psi_{a_i}$, where $(\psi_{a_i}, \psi_{a'_j}) = \delta_{aa'} \delta_{ij}$.

The coefficient α_{a_i} is obtained by taking the inner product

$$(\psi_{a_i}, \psi) = \alpha_{a_i}. \tag{1.28}$$

The formulation of the 2nd postulate of quantum mechanics is purely artificial. In fact, it has been proved and well studied that a function space can be spanned by the eigenvectors of a Hermitian Sturm-Liouville operator. Since the dynamical operators in quantum system are not confined to those of the Sturm-Liouville form, we would formulate on purpose the second postulate of quantum mechanics in order to build and integrate the whole mathematical structure and the logical development of the quantum theory on a solid and self consistent ground.

1.6 Commutability and compatibility of dynamical observables

We shall introduce in the following subsections some terminologies which will help to differentiate various types of the compatible observable.

1.6.1 Compatible observables

If there exist a complete set of linearly independent vectors ψ_{a_i} which are eigenstates of both operators **R** and **S**, then the two physical observables corresponding respectively to Hermitian operators **R** and **S** are said to be **compatible**.

> ### *Proposition 4.*
>
> *If two observables are compatible, their corresponding operators* **R** *and* **S** *commute, i.e.* $[\mathbf{R}, \mathbf{S}] = 0$.

It is obvious because **R** and **S** have the following properties:

$$\mathbf{R}\psi_a = r_a\psi_a, \quad \text{and} \quad \mathbf{S}\psi_a = s_a\psi_a,$$

which lead to

$$(\mathbf{RS} - \mathbf{SR})\psi_a = [\mathbf{R}, \mathbf{S}]\psi_a = 0,$$

if we define the **commutator** of **R** and **S** as $[\mathbf{R}, \mathbf{S}] = \mathbf{RS} - \mathbf{SR}$. Therefore $[\mathbf{R}, \mathbf{S}]\psi = 0$ for any ψ, and Proposition 4 is established.

> ### *Proposition 5.*
>
> *If* **R** *and* **S** *are operators corresponding to two compatible observables, and if* ψ_r *are eigenvectors of* **R***, then*
>
> $$(\psi_r, \mathbf{S}\psi_{r'}) = 0, \quad \text{for} \quad r \neq r'. \tag{1.29}$$

The proof of Proposition 5 is straightforward, i.e.

$$r'(\psi_r, \mathbf{S}\psi_{r'}) = (\psi_r, \mathbf{SR}\psi_{r'}) = (\psi_r, \mathbf{RS}\psi_{r'}) = (\mathbf{R}\psi_r, \mathbf{S}\psi_{r'}) = r(\psi_r, \mathbf{S}\psi_{r'}).$$

We have $(r' - r)(\psi_r, \mathbf{S}\psi_{r'}) = 0$. Hence Proposition 5 is proved.

> ### *Proposition 6.*
>
> *If* $\mathbf{P}_r$ *is the projection operator onto subspace with vectors* ψ_r, *then*
> $$[\mathbf{P}_r, \mathbf{S}] = 0.$$

The proof is again straightforward. For

$$\mathbf{P}_r \psi_{r'} = \delta_{rr'} \psi_{r'}, \tag{1.30}$$

we have

$$
\begin{aligned}
(\psi_{r'}, [\mathbf{P}_r, \mathbf{S}] \psi_{r''}) &= (\psi_{r'}, (\mathbf{P}_r \mathbf{S} - \mathbf{S}\mathbf{P}_r) \psi_{r''}) \\
&= (\mathbf{P}_r \psi_{r'}, \mathbf{S}\psi_{r''}) - (\psi_{r'}, \mathbf{S}\mathbf{P}_r \psi_{r''}) \\
&= (\delta_{rr'} - \delta_{rr''})(\psi_{r'}, \mathbf{S}\psi_{r''}) \equiv 0, \quad \text{(prove it)}
\end{aligned}
$$

that leads to $(\psi, [\mathbf{P}_r, \mathbf{S}] \psi) = 0$ for all ψ, hence

$$[\mathbf{P}_r, \mathbf{S}] = 0. \tag{1.31}$$

Proposition 7.

If $\mathbf{R}$ *and* $\mathbf{S}$ *are two commuting Hermitian operators, there exists a complete set of states which are simultaneously eigenvectors of* $\mathbf{R}$ *and* $\mathbf{S}$.

Let us construct a vector $\phi_r^{(s)}$, projected by $\mathbf{P}_s$ upon the vector ψ_r which is the eigenvector of the Hermitian operator $\mathbf{R}$, i.e.

$$\phi_r^{(s)} = \mathbf{P}_s \psi_r. \tag{1.32}$$

It is obvious that $\phi_r^{(s)}$ is automatically the eigenvector of the Hermitian operator $\mathbf{S}$ with eigenvalue s, namely

$$\mathbf{S}\phi_r^{(s)} = s\phi_r^{(s)}. \tag{1.33}$$

On the other hand, Proposition 6 ensures that $\phi_r^{(s)}$ is also an eigenvector of the operator $\mathbf{R}$, because

$$\mathbf{R}\phi_r^{(s)} = \mathbf{R}\mathbf{P}_s \psi_r = \mathbf{P}_s \mathbf{R} \psi_r = r\mathbf{P}_s \psi_r = r\phi_r^{(s)}. \tag{1.34}$$

Hence Proposition 7 is proved.

1.6.2 Intrinsic compatibility of the dynamical observables and the direct product space

We have seen in the last section that if two operators corresponding respectively to two dynamical observables commute with each other, there always exists a complete set of states which are simultaneously eigenvectors of these two operators. The construction of the simultaneous eigenvectors could be simplified even further in some particular cases in which the compatibility of these two operators is solely based upon the first and the second fundamental **commutation relations**, i.e. $[q_i, q_j] = [p_i, p_j] = 0$, without making use of the third fundamental commutation relation. The dynamical observables corresponding to the commuting operators in this particular category are said to be intrinsic compatible. The construction of the simultaneous eigenvectors of the intrinsic compatible observables is formulated in the following proposition.

Proposition 8.

Let $\mathbf{A}$ and $\mathbf{B}$ be two operators corresponding respectively to two intrinsic compatible observables, and let ψ_{a_i} and φ_{b_j} be the eigenvectors of $\mathbf{A}$ and $\mathbf{B}$ with the eigenvalues a_i and b_j respectively, then the direct product of ψ_{a_i} and φ_{b_j}, denoted by $\psi_{a_i} \otimes \varphi_{b_j}$ is the eigenvector of the operator $\mathbf{F}(\mathbf{A}, \mathbf{B})$ with the eigenvalue $F(a_i, b_j)$.

It can be easily shown that $\psi_{a_i} \otimes \varphi_{b_j}$ is the simultaneous eigenvector of $\mathbf{A}$ and $\mathbf{B}$ with eigenvalues a_i and b_j respectively if we make the following identifications:

$$\mathbf{A} = \mathbf{F}(\mathbf{A}, \mathbf{I}), \quad \text{and} \quad \mathbf{B} = \mathbf{F}(\mathbf{I}, \mathbf{B}).$$

Two operators in different Hilbert spaces are always of intrinsic compatibility. Consider the system of a particle with spin, the physical observable associated with the configuration space is compatible with the physical observable in the spin space. Therefore one is able to express the quantum state as the direct product of a vector in the configuration space and another vector in spin space.

We shall leave it for the reader to verify that Propositions 5 through 7 are consistent with the above formulation. We will also elaborate more on the algebra of the direct product space in Section 2.5 in order to provide a rigorous proof for Proposition 8.

1.6.3 3rd postulate of quantum mechanics and commutator algebra

3rd postulate of quantum mechanics:

Every Poisson bracket in classical mechanics for canonical variables (p_i, q_j) is replaced by the commutator of the corresponding operators with the following relations:

Classical mechanics		Quantum mechanics
$[q_i, q_j] = 0$	$\rightarrow$	$[Q_i, Q_j] = 0$
$[p_i, p_j] = 0$	$\rightarrow$	$[P_i, P_j] = 0$
$[p_i, q_j] = \delta_{ij}$	$\rightarrow$	$[P_i, Q_j] = \dfrac{\hbar}{i}\delta_{ij}$

where $h = 2\pi\hbar$ is Planck's constant.

We discuss the commutator algebra for further applications in later chapters. The commutator of operators **A** and **B**, as it is defined previously

$$[\mathbf{A}, \mathbf{B}] = \mathbf{AB} - \mathbf{BA}, \tag{1.35}$$

then we have

$$[\mathbf{A}, \mathbf{B}] = -[\mathbf{B}, \mathbf{A}], \tag{1.36a}$$

$$[\mathbf{A}, \mathbf{A}] = \mathbf{O}, \tag{1.36b}$$

$$[\mathbf{A}, \mathbf{B} + \mathbf{C}] = [\mathbf{A}, \mathbf{B}] + [\mathbf{A}, \mathbf{C}], \tag{1.36c}$$

$$[\mathbf{A}, [\mathbf{B}, \mathbf{C}]] + [\mathbf{B}, [\mathbf{C}, \mathbf{A}]] + [\mathbf{C}, [\mathbf{A}, \mathbf{B}]] = \mathbf{O}. \tag{1.36d}$$

An operator $\mathbf{C}$ is called a **constant operator** if it commutes with any operator corresponding to the dynamical observables. Obviously any real number times unit operator $\mathbf{I}$ is a Hermitian constant operator.

When an operator is exponentiated, namely $e^{\mathbf{A}}$, it is defined as the usual sense of the exponential function, i.e.

$$e^{\mathbf{A}} = \mathbf{I} + \mathbf{A} + \frac{1}{2!}\mathbf{A}^2 + \frac{1}{3!}\mathbf{A}^3 + \dots . \qquad (1.37)$$

We shall now show a useful identity

$$e^{\mathbf{A}}\mathbf{B}e^{-\mathbf{A}} \equiv \mathbf{B} + \frac{1}{1!}[\mathbf{A}, \mathbf{B}] + \frac{1}{2!}[\mathbf{A}, [\mathbf{A}, \mathbf{B}]] + \dots . \qquad (1.38)$$

Proof

Let $f(\lambda) = e^{\lambda \mathbf{A}}\mathbf{B}e^{-\lambda \mathbf{A}}$ and expand the function $f(\lambda)$ in terms of power series of λ at $\lambda = 0$, i.e.

$$f(\lambda) = f(0) + \frac{\lambda}{1!}f'(0) + \frac{\lambda^2}{2!}f''(0) + \dots .$$

Since,

$$f'(\lambda) = \mathbf{A}f(\lambda) - f(\lambda)\mathbf{A} = [\mathbf{A}, f(\lambda)],$$

we can evaluate each order of the derivatives of $f(\lambda)$ at $\lambda = 0$, i.e.

$$f(0) = \mathbf{B},$$
$$f'(0) = [\mathbf{A}, \mathbf{B}],$$
$$f''(0) = [\mathbf{A}, [\mathbf{A}, \mathbf{B}]],$$

and thus

$$f(\lambda) = f(0) + \frac{\lambda}{1!}f'(0) + \frac{\lambda^2}{2!}f''(0) + \dots$$

becomes

$$f(\lambda) = e^{\lambda \mathbf{A}} \mathbf{B} e^{-\lambda \mathbf{A}} = \mathbf{B} + \lambda[\mathbf{A}, \mathbf{B}] + \frac{\lambda^2}{2!}[\mathbf{A}, [\mathbf{A}, \mathbf{B}]] + \dots$$

Therefore we reach the identity by setting $\lambda = 1$ in the function $f(\lambda)$, i.e.

$$f(1) = e^{\mathbf{A}} \mathbf{B} e^{-\mathbf{A}} \equiv \mathbf{B} + [\mathbf{A}, \mathbf{B}] + \frac{1}{2!}[\mathbf{A}, [\mathbf{A}, \mathbf{B}]] + \dots \qquad (1.39)$$

1.7 Non-commuting operators and the uncertainty principle

The uncertainty in the simultaneous measurement of the coordinate and its conjugate momentum is due to the dual nature of a quantum particle, namely the **particle-wave duality**. From the mathematical point of view, the uncertainty originates from the non-commutability of the two operators that correspond respectively to these two physical observables. In fact, uncertainty will arise for any simultaneous measurement of a pair of dynamical quantities if their corresponding quantum operators do not commute. A demonstration is in order to re-establish Heisenberg's famous uncertainty relations. For simplicity, we only consider the one-dimensional case and denote Δq and Δp to be the uncertainty in coordinate and momentum simultaneous measurement respectively. Δq and Δp are also called the variance, or the deviation of q and p correspondingly defined as

$$(\Delta q)^2 = (\psi, (\mathbf{Q} - \langle \mathbf{Q} \rangle \mathbf{I})^2 \psi), \qquad (1.40)$$

$$(\Delta p)^2 = (\psi, (\mathbf{P} - \langle \mathbf{P} \rangle \mathbf{I})^2 \psi), \qquad (1.41)$$

where ψ is the quantum state of interest, $\langle \mathbf{Q} \rangle \mathbf{I}$ and $\langle \mathbf{P} \rangle \mathbf{I}$ are both constant operators obtained by means of multiplying respectively the expectation value $\langle \mathbf{Q} \rangle$ and $\langle \mathbf{P} \rangle$ by a unit operator.

Let us introduce **deviation operators** $\mathbf{Q}_d$ and $\mathbf{P}_d$ as follows:

$$\mathbf{Q}_d = \mathbf{Q} - \langle \mathbf{Q} \rangle \mathbf{I}, \tag{1.42}$$

$$\mathbf{P}_d = \mathbf{P} - \langle \mathbf{P} \rangle \mathbf{I}, \tag{1.43}$$

and denote

$$\phi_1 = \mathbf{Q}_d \psi, \quad (\Delta q)^2 = (\phi_1, \phi_1), \tag{1.44}$$

$$\phi_2 = \mathbf{P}_d \psi, \quad (\Delta p)^2 = (\phi_2, \phi_2). \tag{1.45}$$

Apply the Schwarz inequality for ϕ_1 and ϕ_2, i.e.

$$(\phi_1, \phi_1)(\phi_2, \phi_2) \geqslant |(\phi_1, \phi_2)|^2. \tag{1.46}$$

Thus we have

$$(\Delta q)^2 (\Delta p)^2 \geqslant |(\mathbf{Q}_d \psi, \mathbf{P}_d \psi)|^2 = |(\psi, \mathbf{Q}_d \mathbf{P}_d \psi)|^2. \tag{1.47}$$

Let us introduce an anticommutator of $\mathbf{Q}_d$ and $\mathbf{P}_d$, defined as

$$\{\mathbf{Q}_d, \mathbf{P}_d\} = \mathbf{Q}_d \mathbf{P}_d + \mathbf{P}_d \mathbf{Q}_d. \tag{1.48}$$

It can be readily shown that the expectation value of an anticommutator with respect to the state ψ is always real, say $\alpha = (\psi, \{\mathbf{Q}_d, \mathbf{P}_d\}\psi)$. Furthermore since $\mathbf{Q}_d \mathbf{P}_d = \frac{1}{2}[\mathbf{Q}_d, \mathbf{P}_d] + \frac{1}{2}\{\mathbf{Q}_d, \mathbf{P}_d\}$, and its expectation value is calculated as

$$(\psi, \mathbf{Q}_d \mathbf{P}_d \psi) = \frac{1}{2}(\psi, [\mathbf{Q}_d, \mathbf{P}_d]\psi) + \frac{1}{2}(\psi, \{\mathbf{Q}_d, \mathbf{P}_d\}) = \frac{i}{2}\hbar + \frac{1}{2}\alpha. \tag{1.49}$$

Hence **Heisenberg's uncertainty** relation in the one dimensional case is derived, i.e.

$$(\Delta q)^2 (\Delta p)^2 \geqslant \frac{\hbar^2}{4} + \frac{\alpha^2}{4} \geqslant \frac{\hbar^2}{4}, \quad \text{or} \quad \Delta q \Delta p \geqslant \frac{1}{2}\hbar. \tag{1.50}$$

The minimum uncertainty relation holds if the following conditions are fulfilled:

(a) $\mathbf{P}_d \psi = \lambda \mathbf{Q}_d \psi$, namely $\mathbf{P}_d \psi$ and $\mathbf{Q}_d \psi$ are linearly dependent.

(b) $\alpha = 0$.

The coefficient λ can be evaluated by taking the expectation value of the anticommutator $\frac{1}{2}\{\mathbf{Q}_d, \mathbf{P}_d\}$ with respect to the state ψ, and putting it equal to zero. We reach

$$\lambda(\Delta q)^2 + \frac{1}{\lambda}(\Delta p)^2 = 0.$$

On the other hand, the expectation value of the commutator $[\mathbf{Q}_d, \mathbf{P}_d] = \frac{\hbar}{i}$ leads to $\lambda(\Delta q)^2 - \frac{1}{\lambda}(\Delta p)^2 = \frac{\hbar}{i}$, hence

$$\lambda = \frac{i\hbar}{2(\Delta q)^2}. \tag{1.51}$$

An explicit expression of the state function in terms of the least uncertainty Δq shall be postponed until we cover the materials on q-representation in quantum mechanics.

1.8 Exercises

Ex 1.8.1

Consider the following set of continuous functions $f_n(x) = x^n, x \in [-1, 1]$ which spans a $\mathcal{L}^2(-1, 1)$-space. Find explicitly the first 3 orthonormal functions by the Gram-Schmidt process. What are those functions that occur to your mind?

Ex 1.8.2

Prove that the Minkowski inequality holds in Hilbert space, i.e.

$$\|\psi + \phi\| \leqslant \|\psi\| + \|\phi\|.$$

(**Hint:** take the square of either side.)

Ex **1.8.3**

Prove the law of parallelogram holds in Hilbert space, i.e.

$$\|\psi + \phi\|^2 + \|\psi - \phi\|^2 = 2(\|\psi\|^2 + \|\phi\|^2).$$

Ex **1.8.4**

Prove that every finite dimensional vector space is complete. (**Hint:** since the real and complex numbers are complete.)

Ex **1.8.5**

Show that the $\|\mathbf{AB}\| \leqslant \|\mathbf{A}\|\|\mathbf{B}\|$ if $\mathbf{A}, \mathbf{B} \in$ bounded operator.

Ex **1.8.6**

Show that the expectation value of any dynamical observable in the physical system is always real.

Ex **1.8.7**

Prove that the adjoint conjugate can equivalently be defined as $(\mathbf{A}\psi, \phi) = (\psi, \mathbf{A}^\dagger \phi)$ if $(\mathbf{A}^\dagger \psi, \phi) = (\psi, \mathbf{A}\phi)$.

Ex **1.8.8**

Let $\{\phi_1, \phi_2, \ldots, \phi_n\}$ be a set of an orthonormal basis. Prove that operator $\mathbf{U}$ is unitary if $\{\mathbf{U}\phi_1, \mathbf{U}\phi_2, \ldots, \mathbf{U}\phi_n\}$ is also a set of an orthonormal basis.

Ex **1.8.9**

Prove that if $\mathbf{A}$ and $\mathbf{B}$ are two operators that both commute with their commutator, then

$$e^{\mathbf{A}} e^{\mathbf{B}} = e^{\mathbf{A}+\mathbf{B}+[\mathbf{A},\mathbf{B}]/2}.$$

(**Hint:** let $f(\lambda) = e^{\lambda A} e^{\lambda B} e^{-\lambda(A+B)}$, and obtain $\frac{df(\lambda)}{d\lambda} = \lambda[A, B]f(\lambda)$ by Eq. (1.39). Then integrate it.)

Ex **1.8.10**

Consider a Hilbert space spanned by a Hermitian operator $\mathbf{A}$.

(a) Prove that $\prod_a (\mathbf{A} - a\mathbf{I})$ is a null operator if $\mathbf{A}\psi_a = a\psi_a$.

(b) What is the significance of the following operator, $\displaystyle\prod_{a \neq a'} \frac{\mathbf{A} - a\mathbf{I}}{a' - a}$?

Ex **1.8.11**

If two observables $\mathbf{A}_1$ and $\mathbf{A}_2$ are not compatible, but their corresponding operators both commute with the Hamiltonian operator $\mathbf{H}$, i.e.

$$[\mathbf{A}_1, \mathbf{H}] = [\mathbf{A}_2, \mathbf{H}] = 0.$$

Show that the energy eigenstates are in general degenerate.

Ex **1.8.12**

Consider a one-dimensional Hamiltonian

$$\mathbf{H} = \frac{1}{2m}\mathbf{P}^2 + V(\mathbf{Q})$$

and use the fact that the commutator of $\mathbf{Q}$ and $[\mathbf{Q}, \mathbf{H}]$ is a constant operator, to show that

$$\sum_k (E_k - E_s)|Q_{sk}|^2 = \frac{\hbar^2}{2m},$$

which is referred to as the Thomas-Reiche-Kuhn sum rule, where $Q_{sk} = (\psi_s, \mathbf{Q}\psi_k)$ and ψ_s is the eigenstate of $\mathbf{H}$ with eigenvalue E_s, i.e. $\mathbf{H}\psi_s = E_s\psi_s$.

Chapter 2

Space-Time Translation, Quantum Dynamics and Various Representations in Quantum Mechanics

2.1 Vector space and dual vector space

We start from $\mathcal{C}^n$-space as an example. In matrix notation, a vector $\bar{x}$ in $\mathcal{C}^n$-space can be represented by a vertical array of n complex numbers as follows

$$\bar{x} = \begin{pmatrix} x_1 \\ x_2 \\ \vdots \\ x_i \\ x_{i+1} \\ \vdots \\ x_n \end{pmatrix}, \tag{2.1}$$

which is called a column matrix, where x_i is the ith-component of the vector. Each component is labeled by a subscript i. Matrix algebra ensures that the two basic operations of vectors, as we explained at the beginning of the first chapter, can be reproduced. The orthonormal basis of the n-dimensional vector space $\bar{e}_i$ is also represented by a column matrix. Take the ith-basis as an example:

$$
\bar{e}_i = \begin{pmatrix} 0 \\ 0 \\ \vdots \\ 1 \\ 0 \\ \vdots \\ 0 \end{pmatrix}, \tag{2.2}
$$

that allows one to express the vector $\bar{x}$ as the following linear combination

$$
\bar{x} = \sum_{i=1}^{n} x_i \bar{e}_i. \tag{2.3}
$$

To perform the inner product of an ordered pair of vectors in $\mathcal{C}^n$, one needs to introduce another vector space $^*\mathcal{C}^n$, a **dual space** to $\mathcal{C}^n$-space. A vector in $^*\mathcal{C}^n$ is denoted by a row matrix as follows

$$
\tilde{x} = \begin{pmatrix} x_1^* & x_2^* & \cdots & x_i^* & x_{i+1}^* & \cdots & x_n^* \end{pmatrix},
$$

which is exactly the adjoint conjugate, also called the **Hermitian conjugate** of vector $\bar{x}$ in $\mathcal{C}^n$-space, namely by taking the complex conjugate of each entry and transposing the column matrix into a row matrix in $^*\mathcal{C}^n$-space. One justifies immediately that $^*\mathcal{C}^n$ is also a vector space.

Let us see how to perform an inner product by means of matrix algebra. Take the ordered pair of vectors $\bar{x}$ and $\bar{y}$ as an example. We construct a dual vector $\tilde{x}$, which is an adjoint conjugate of the first vector $\bar{x}$. Then the inner product of vectors $\bar{x}$ and $\bar{y}$ is obtained from the multiplication of the row matrix $\tilde{x}$ in $^*\mathcal{C}^n$-space by the column matrix $\bar{y}$ in $\mathcal{C}^n$-space, i.e.

$$\tilde{x}\bar{y} = \begin{pmatrix} x_1^* & x_2^* & \cdots & x_n^* \end{pmatrix} \begin{pmatrix} y_1 \\ y_2 \\ \vdots \\ y_n \end{pmatrix}. \tag{2.4}$$

The linear operator $\mathbf{A}$ in $\mathcal{C}^n$-space is represented by a square matrix with matrix elements A_{ij} for $i, j = 1, 2, \ldots, n$, obtained by sandwiching the operator $\mathbf{A}$ between the basis row vector $\tilde{e}_i$ and the basis column vector $\bar{e}_j$, i.e.

$$\tilde{e}_i \mathbf{A} \bar{e}_j = A_{ij}. \tag{2.5}$$

If we take the adjoint conjugate of the above equation, we have

$$\tilde{e}_j \mathbf{A}^\dagger \bar{e}_i = (A^\dagger)_{ji} \tag{2.6}$$

on the left hand side of Eq. (2.5), while on the right hand side of Eq. (2.5), we have merely a complex number, because its adjoint conjugate is just the complex conjugate. Hence we have the following relation

$$(A^\dagger)_{ji} = A_{ij}^*. \tag{2.7}$$

A matrix $\mathbf{A}$ which equals to its adjoint conjugate matrix, namely $\mathbf{A} = \mathbf{A}^\dagger$ is called a Hermitian matrix. Then we have

$$A_{ij} = (A^\dagger)_{ij} = A_{ji}^*, \tag{2.8}$$

or $A_{ij}^* = A_{ji}$, by taking the complex conjugate on both sides.

The expectation value of an operator $\mathbf{A}$ with respect to the vector $\bar{x}$ is taken by multiplying the row vector $\tilde{x}$, the n by n matrix $\mathbf{A}$, and the column vector $\bar{x}$ successively as follows

$$\tilde{x} \mathbf{A} \bar{x} = \text{expetation value of } \mathbf{A} \text{ with respect to } \bar{x}.$$

If the operator stands for a physical observable, then it is a Hermitian matrix and the expectation value is real because

$$\tilde{x} \mathbf{A} \bar{x} = \tilde{x} \mathbf{A}^\dagger \bar{x} = (\tilde{x} \mathbf{A} \bar{x})^\dagger = (\tilde{x} \mathbf{A} \bar{x})^*. \tag{2.9}$$

The matrix algebra also ensures that $n \times n$ square matrices follow the rules of being linear operators.

Matrix representation for C^n-space allows us to construct the projection operator readily. A projection operator that projects a vector $\bar{x}$ onto a subspace spanned by basis $\bar{e}_i$ is a square matrix formed by putting a column basis before a row basis, i.e.

$$\mathbf{P}_i = \bar{e}_i \tilde{e}_i, \qquad (2.10)$$

or more specifically,

$$\mathbf{P}_i = \begin{pmatrix} 0 \\ 0 \\ \vdots \\ 1 \\ 0 \\ \vdots \\ 0 \end{pmatrix} \begin{pmatrix} 0 & 0 & \cdots & 1 & 0 & \cdots & 0 \end{pmatrix} = \begin{pmatrix} 0 & 0 & \cdots & 0 & 0 & \cdots & 0 \\ 0 & \ddots & & & & & \vdots \\ \vdots & & \ddots & & & & \vdots \\ 0 & & & 1 & & & 0 \\ 0 & & & & 0 & & 0 \\ \vdots & & & & & \ddots & \vdots \\ 0 & 0 & \cdots & 0 & 0 & \cdots & 0 \end{pmatrix}.$$

The projection operator $\mathbf{P}_{\mathcal{M}}$ onto a subspace $\mathcal{M}$ simply takes the sum of the projection operators $\sum \mathbf{P}_i$, i.e.

$$\mathbf{P}_{\mathcal{M}} = \sum \mathbf{P}_i \quad \text{for} \quad e_i \in \mathcal{M}. \qquad (2.11)$$

It is obvious that the sum of the projection operators $\sum_{i=1}^{n} \mathbf{P}_i$ over the basis in the entire space is an identity operator, i.e.

$$\sum_{i=1}^{n} \mathbf{P}_i = \mathbf{I}. \qquad (2.12)$$

It is advantageous in this matrix representation that it allows one to easily construct any projection operator. If one wants to project a vector onto a particular vector $\bar{y}$, the corresponding projection operator can then be expressed as

$$\mathbf{P}_y = \bar{y}\tilde{y}. \tag{2.13}$$

The Hilbert space corresponding to a physical system is often infinite dimensional and also not necessarily separable. It becomes awkward or even impossible to use the matrix formulation in those cases.

To overcome the inconvenience and the difficulty, Dirac introduced new notations for the vectors in Hilbert space. A vector in $\mathcal{H}$ is denoted by the symbol $|\alpha\rangle$, called a **ket vector**, or simply a ket. While a vector in the adjoint conjugate dual space $^*\mathcal{H}$ is denoted by the symbol $\langle\alpha|$, called a **bra vector**, or simply a bra. Similar to the case in $\mathcal{C}^n$-space, to construct the inner product of an ordered pair of vectors in Dirac notation, say the vectors $|\alpha\rangle$ and $|\beta\rangle$, we put the adjoint conjugate of $|\alpha\rangle$, namely the bra vector $\langle\alpha|$ in front of the ket vector $|\beta\rangle$ as follows

$$\langle\alpha||\beta\rangle \equiv \langle\alpha|\beta\rangle, \tag{2.14}$$

which reaffirms the definition of the adjoint conjugates of operator $\mathbf{A}$.

The square norm of a vector $|\alpha\rangle$ is defined as the inner product of a pair of same vectors, i.e.

$$\||\alpha\rangle\|^2 = \langle\alpha|\alpha\rangle. \tag{2.15}$$

A linear operator $\mathbf{A}$ which transforms a ket vector $|\alpha\rangle$ into another ket vector $|\alpha'\rangle$ is written as

$$|\alpha'\rangle = \mathbf{A}|\alpha\rangle. \tag{2.16}$$

The inner product of the ordered pair of vectors of $|\alpha\rangle$ and $|\beta'\rangle$ is written as follows

$$\langle\alpha|\beta'\rangle = \langle\alpha|\,(\mathbf{A}|\beta\rangle) \equiv \langle\alpha|\mathbf{A}|\beta\rangle. \tag{2.17}$$

Generally we omit writing the parenthesis as it appears in the second term of last equation.

On the other hand, if the inner product is constructed for an ordered pair of vectors $|\alpha'\rangle$ and $|\beta\rangle$, in which the vector $|\alpha'\rangle$ is a transformed vector of $|\alpha\rangle$ by operator $\mathbf{A}$, then the corresponding inner product is written as

$$\langle \alpha' | \beta \rangle = \langle \alpha | \mathbf{A}^\dagger | \beta \rangle. \tag{2.18}$$

Similar to the case in $\mathcal{C}^n$-space, we conclude that

$$\langle \alpha | \mathbf{A}^\dagger | \beta \rangle = \langle \beta | \alpha' \rangle^* = \langle \beta | \mathbf{A} | \alpha \rangle^*, \tag{2.19}$$

and the definition of adjoint conjugate of $\mathbf{A}$ is regained.

Dirac notations in vector space have the advantage of constructing the projection operator readily. Consider a Hermitian operator $\mathbf{A}$, and the eigenvector $|a\rangle$ with the corresponding eigenvalue a, i.e.

$$\mathbf{A}|a\rangle = a|a\rangle. \tag{2.20}$$

The set of eigenvectors $\{|a_1\rangle, |a_2\rangle, \ldots, |a_i\rangle, \ldots\}$ forms the basis of the Hilbert space $\mathcal{H}$.

The projection operator onto the subspace $|\alpha\rangle$ is designed by putting a ket vector and a bra vector together in successive order, i.e.

$$\mathbf{P}_\alpha = |\alpha\rangle\langle\alpha|. \tag{2.21}$$

Any vector $|\alpha\rangle$ in the Hilbert space is expressed in a linear combination of its basis as follows

$$|\alpha\rangle = \sum_j \alpha_j |a_j\rangle, \tag{2.22}$$

where the coefficient α_j is obtained through the inner product, i.e.

$$\langle a_i | \alpha \rangle = \langle a_i | \sum_j \alpha_j |a_j\rangle = \sum_j \alpha_j \delta_{ji} = \alpha_i. \tag{2.23}$$

Projection of the vector onto the subspace spanned by the basis $|a_i\rangle$ is then read as

$$\mathbf{P}_i = |a_i\rangle\langle a_i|. \tag{2.24}$$

A projection operator onto subspace $\mathcal{M}$ can also be constructed by

$$\mathbf{P}_\mathcal{M} = \sum_i \mathbf{P}_i = \sum_i |a_i\rangle\langle a_i|, \quad |a_i\rangle \in \mathcal{M}. \tag{2.25}$$

By extending subspace $\mathcal{M}$ to cover the whole Hilbert space, we reach the **closure relation**,

$$\sum \mathbf{P}_i = \sum |a_i\rangle\langle a_i| = \mathbf{I}. \qquad (2.26)$$

In the case of a separable Hilbert space with degeneracy, and denoting the basis by $|a, i\rangle$, the closure relation reads as

$$\sum_{a,i} \mathbf{P}_{a,i} = \sum_{a,i} |a, i\rangle\langle a, i| = \mathbf{I}. \qquad (2.27)$$

For the case that the eigenvalues become continuous, the eigenvectors $|\xi\rangle$ are no longer denumerable. The basis vector is characterized by the eigenvalue ξ, which is a continuum and the inner product is normalized to be a **delta function**, i.e.

$$\langle\xi|\xi'\rangle = \delta(\xi - \xi'), \qquad (2.28)$$

and the closure relation in this case takes the form as

$$\int \mathbf{P}_\xi d\xi = \int |\xi\rangle d\xi \langle\xi| = \mathbf{I}. \qquad (2.29)$$

2.2 q-representation and p-representation in quantum mechanics

Consider the Hilbert space spanned by the eigenvectors of the Hermitian operator $\mathbf{X}$ corresponding to a dynamical observable in coordinate x. For simplicity, we limit ourselves to the one-dimensional case. The eigenvalue equation reads as

$$\mathbf{X}|x\rangle = x|x\rangle. \qquad (2.30)$$

If we make a transport of the physical system to the right by a distance ξ, namely the spatial translation of the system from x to $x + \xi$, it is interesting to observe the change of the eigenvector in the new system.

Let us first introduce a unitary operator, called the **translational operator**, defined as

$$\mathbf{U}(\mathbf{P};\xi) = e^{-\frac{i}{\hbar}\xi\mathbf{P}}, \tag{2.31}$$

where

ξ = real parameter, namely the distance of the spatial translation,

$\mathbf{P}$ = momentum operator,

and $\mathbf{U}(\mathbf{P};\xi)$ is unitary because

$$\mathbf{U}^{\dagger}(\mathbf{P};\xi) = e^{\frac{i}{\hbar}\xi\mathbf{P}} = \mathbf{U}^{-1}(\mathbf{P};\xi).$$

Proposition 1.

$\mathbf{U}(\mathbf{P};\xi)|x\rangle$ *is an eigenvector of operator* $\mathbf{X}$ *with eigenvalue* $x+\xi$.

To prove the above statement, let us start with the commutation relation of $\mathbf{X}$ and $\mathbf{P}$, i.e. $[\mathbf{X},\mathbf{P}] = i\hbar\mathbf{I}$, that enables us to evaluate the following commutators

$$[\mathbf{X},\mathbf{P}^{n}] = \mathbf{P}^{n-1}[\mathbf{X},\mathbf{P}] + [\mathbf{X},\mathbf{P}^{n-1}]\mathbf{P} = in\hbar\mathbf{P}^{n-1}, \tag{2.32}$$

and

$$[\mathbf{X},\mathbf{U}] = i\hbar\left[\mathbf{X},\sum_{m}\frac{1}{m!}\left(-\frac{i}{\hbar}\xi\right)^{m}\mathbf{P}^{m}\right]$$

$$= \xi\sum_{m}\frac{1}{(m-1)!}\left(-\frac{i}{\hbar}\xi\right)^{m-1}\mathbf{P}^{m-1} = \xi\mathbf{U}, \tag{2.33}$$

or

$$\mathbf{X}\mathbf{U} = \mathbf{U}(\mathbf{X} + \xi\mathbf{I}). \tag{2.34}$$

Construct the new state

$$|\alpha\rangle = \mathbf{U}|x\rangle.$$

Since

$$\mathbf{X}|\alpha\rangle = \mathbf{XU}|x\rangle = \mathbf{U}(\mathbf{X} + \xi\mathbf{I})|x\rangle = \mathbf{U}(x + \xi)|x\rangle = (x + \xi)|\alpha\rangle, \quad (2.35)$$

it implies that $|\alpha\rangle$ is also the eigenvector of operator $\mathbf{X}$, with eigenvalue $x + \xi$. Therefore we cast vector $|\alpha\rangle$ as

$$|\alpha\rangle = c|x + \xi\rangle = |x + \xi\rangle.$$

The constant c is just a pure phase factor that is chosen to be a unit without losing its generality.

Proposition 2.

A state function or wave function as it is generally called, is defined as the inner product of the vector $|x\rangle$ and the state vector $|\psi\rangle$ of interest, i.e. $\psi(x) = \langle x|\psi\rangle$.

Furthermore, sandwiching the momentum operator in between bra vector $\langle x|$ and state vector $|\psi\rangle$ results in differentiating the wave function, or more precisely,

$$\langle x|\mathbf{P}|\psi\rangle = \frac{\hbar}{i}\frac{\partial}{\partial x}\psi(x).$$

We consider an infinitesimal translation ξ, then

$$\mathbf{U}(\mathbf{P};\xi) = 1 - \frac{i}{\hbar}\xi\mathbf{P},$$

and

$$\langle x|\mathbf{U}(\mathbf{P};\xi)|x'\rangle = \langle x|x'+\xi\rangle = \langle x|\left(1 - \frac{i}{\hbar}\xi\mathbf{P}\right)|x'\rangle,$$

$$\delta(x - x') - \delta(x - x' - \xi) = \frac{i}{\hbar}\xi\langle x|\mathbf{P}|x'\rangle,$$

$$\langle x|\mathbf{P}|x'\rangle = \frac{\hbar}{i}\frac{d}{dx}\delta(x - x') = \frac{\hbar}{i}\delta'(x - x'),$$

where $\delta'(x - x')$ is the derivative of the delta function with respect to its argument, i.e.

$$\delta'(x - x') = \frac{d}{dx}\delta(x - x') = \frac{d}{d(x - x')}\delta(x - x'),$$

and it has the following property:

$$\int f(x)\delta'(x - a)dx = -f'(a) \tag{2.36}$$

for any function $f(x)$. With this property, we are able to derive that

$$\langle x|\mathbf{P}|\psi\rangle = \int\langle x|\mathbf{P}|x'\rangle dx'\langle x'|\psi\rangle = \int \frac{\hbar}{i}\delta'(x - x')\psi(x')dx'$$

$$= -\frac{\hbar}{i}\int \frac{d}{dx'}\delta(x' - x)\psi(x')dx' = \frac{\hbar}{i}\frac{d}{dx}\psi(x). \tag{2.37}$$

We are now in the position to make a comparison between the abstract state and the corresponding q-representation.

Hilbert space	q-representation
$\lvert\psi\rangle$	$\langle x\lvert\psi\rangle = \psi(x)$
$\mathbf{X}\lvert\psi\rangle$	$\langle x\lvert\mathbf{X}\lvert\psi\rangle = x\psi(x)$
$\mathbf{P}\lvert\psi\rangle$	$\langle x\lvert\mathbf{P}\lvert\psi\rangle = \dfrac{\hbar}{i}\dfrac{d}{dx}\psi(x)$
$F(\mathbf{P},\mathbf{X})\lvert\psi\rangle$	$F(\dfrac{\hbar}{i}\dfrac{d}{dx},x)\psi(x)$

$\psi(x)$ in the above table is called the **wave function** or the **state function**, and the absolute square of $\psi(x)$, i.e. $\lvert\psi(x)\rvert^2$ is commonly referred to as the probability density of the system.

We particularly consider the eigenvalue equation for the energy of the system. The Hamiltonian operator reads

$$\mathbf{H} = \frac{1}{2m}\mathbf{P} + V(\mathbf{X}), \tag{2.38}$$

and the eigenvalue equation

$$\mathbf{H}\lvert\psi\rangle = E\lvert\psi\rangle, \tag{2.39}$$

when it is expressed in terms of q-representation, becomes

$$\langle x\lvert\mathbf{H}\lvert\psi\rangle = E\langle x\lvert\psi\rangle, \tag{2.40}$$

or explicitly, the time independent Schrödinger equation as

$$\left(-\frac{\hbar^2}{2m}\frac{d^2}{dx^2} + V(x)\right)\psi(x) = E\psi(x). \tag{2.41}$$

The p-representation of the state function can be formulated parallel to that of the q-representation with the help of the following proposition.

Proposition 3.

Q-representation of the eigenstate of the momentum operator is a plane wave, namely

$$\langle x|p\rangle = \frac{1}{\sqrt{2\pi\hbar}}e^{\frac{i}{\hbar}px}$$

where $|p\rangle$ is the eigenstate of $\mathbf{P}$, i.e, with the eigenvalue p, i.e. with $\mathbf{P}|p\rangle = p|p\rangle$.

Let us consider $\langle x|\mathbf{P}|p\rangle$ and insert the identity operator $\int |x'\rangle dx'\langle x'|$ after the momentum operator $\mathbf{P}$. We have

$$\langle x|\mathbf{P}|p\rangle = \int \langle x|\mathbf{P}|x'\rangle dx'\langle x'|p\rangle = \int \frac{\hbar}{i}\delta'(x-x')\langle x'|p\rangle dx' = \frac{\hbar}{i}\frac{d}{dx}\langle x|p\rangle,$$

or

$$p\langle x|p\rangle = \frac{\hbar}{i}\frac{d}{dx}\langle x|p\rangle.$$

Solving this differential equation, we obtain

$$\langle x|p\rangle = Ae^{\frac{i}{\hbar}px} = \frac{1}{\sqrt{2\pi\hbar}}e^{\frac{i}{\hbar}px}. \tag{2.42}$$

The normalization constant can be verified as follows,

$$\langle x|x'\rangle = \delta(x-x') = \int \langle x|p\rangle dp\langle p|x'\rangle$$

$$= |A|^2 \int e^{\frac{i}{\hbar}p(x-x')}dp = 2\pi\hbar|A|^2\delta(x-x').$$

That implies $|A|^2 = 1/2\pi\hbar$.

Similar to the case in the q-representation, the wave function in p-space reads as

$$\langle p|\psi\rangle = \psi(p). \tag{2.43}$$

In fact $\psi(p)$ is the inverse Fourier transform of $\psi(x)$, i.e.

$$\psi(p) = \langle p|\psi\rangle = \int \langle p|x'\rangle dx' \langle x'|\psi\rangle = \frac{1}{\sqrt{2\pi\hbar}} \int e^{\frac{i}{\hbar}px'} dx' \psi(x'). \tag{2.44}$$

The Schrödinger equation in the p-representation can be written as

$$\langle p|\mathbf{H}|\psi\rangle = \langle p|\frac{1}{2m}\mathbf{P}^2 + V(\mathbf{X})|\psi\rangle. \tag{2.45}$$

The first term is trivially obtained,

$$\langle p|\frac{1}{2m}\mathbf{P}^2|\psi\rangle = \frac{1}{2m}p^2\psi(p) \tag{2.46}$$

while the second term can be evaluated by the closure relations,

$$\langle p|V(\mathbf{X})|\psi\rangle = \int \langle p|V(\mathbf{X})|p'\rangle dp' \langle p'|\psi\rangle,$$

where

$$\langle p|V(\mathbf{X})|p'\rangle = \iint \langle p|x'\rangle dx' \langle x'|V(\mathbf{X})|x''\rangle dx'' \langle x''|p'\rangle$$

$$= \frac{1}{2\pi\hbar} \iint e^{\frac{i}{\hbar}(p'x''-px')} V(x')\delta(x'-x'')dx'dx''$$

$$= \frac{1}{2\pi\hbar} \int e^{\frac{i}{\hbar}(p'-p)x} V(x)dx = \tilde{V}(p-p'), \tag{2.47}$$

is the Fourier transform of the potential in p-space. Finally we reach the Schrödinger equation in the p-representation as follows

$$\frac{p^2}{2m}\psi(p) + \int \tilde{V}(p - p')\psi(p')dp' = E\psi(p). \tag{2.48}$$

Before leaving this section, let us go back to the q-representation of the state with the minimum quantum uncertainty $\Delta p \Delta q = \hbar/2$ discussed in Section 1.7, where $|\psi\rangle$ satisfies the following condition

$$\boldsymbol{P}_d|\psi\rangle = \frac{i\hbar}{2(\Delta q)^2}\boldsymbol{Q}_d|\psi\rangle. \tag{2.49}$$

The q-representation of the above equation reads as

$$\left(\frac{\hbar}{i}\frac{\partial}{\partial x} - \langle p \rangle\right)\psi(x) = \frac{i\hbar}{2(\Delta x)^2}(x - \langle x \rangle)\psi(x), \tag{2.50}$$

which allows us to obtain the normalized wave function of minimum uncertainty as

$$\psi(x) = \frac{1}{[2\pi(\Delta x)^2]^{1/4}} \exp\left[\frac{(x - \langle x \rangle)^2}{4(\Delta x)^2} + \frac{i\langle p \rangle x}{\hbar}\right]. \tag{2.51}$$

2.3 Harmonic oscillator revisited

There exists very rich structure in dealing with the harmonic oscillation mathematically on a quantum level. We shall attack the problem by means of the number operator through the **creation operator** and the **annihilation operator** rather than solving the Schrödinger equation in the q-representation, in which the quantization of the energy levels is derived from the boundary conditions of the wave functions.

The technique developed in this section can even be applied to the system of linear oscillators in the limit of continuum, which gives rise to the quantization of the field.

2.3.1 Creation and annihilation operators

The Hamiltonian operator for the one-dimensional harmonic oscillator reads as

$$\mathbf{H} = \frac{1}{2m}\mathbf{P}^2 + \frac{1}{2}m\omega^2\mathbf{X}^2. \tag{2.52}$$

Let us introduce the new variables defined as follows:

$$p_\xi = \frac{1}{\sqrt{m\omega\hbar}}\mathbf{P}, \quad \xi = \sqrt{\frac{m\omega}{\hbar}}\mathbf{X}. \tag{2.53}$$

The Hamiltonian operator takes simpler form,

$$\mathbf{H} = \hbar\omega\left(\frac{1}{2}p_\xi^2 + \frac{1}{2}\xi^2\right) = \hbar\omega\mathcal{H}, \tag{2.54}$$

with $\mathcal{H}$ defined as $\mathcal{H} = (p_\xi^2 + \xi^2)/2$.

The nonvanishing canonical quantization relation of $\mathbf{P}$ and $\mathbf{X}$, i.e. $[\mathbf{P}, \mathbf{X}] = \hbar\mathbf{I}/i$ can also be simplified as

$$[p_\xi, \xi] = -i\mathbf{I}. \tag{2.55}$$

Define the annihilation operator and the creation operator as follows

$$a = \frac{1}{\sqrt{2}}(\xi + ip_\xi), \quad \text{and} \quad a^\dagger = \frac{1}{\sqrt{2}}(\xi - ip_\xi), \tag{2.56}$$

or reversely

$$\xi = \frac{1}{\sqrt{2}}(a^\dagger + a), \tag{2.57}$$

$$p_\xi = \frac{i}{\sqrt{2}}(a^\dagger - a). \tag{2.58}$$

The commutation relations become:

$$[a, a] = [a^\dagger, a^\dagger] = \mathbf{O}, \qquad [a, a^\dagger] = \mathbf{I}. \tag{2.59}$$

Then $\mathcal{H}$ can be cast into $\mathcal{H} = (a^\dagger a + a a^\dagger)/2 = N + \mathbf{I}/2$ where $N = a^\dagger a$, a Hermitian operator called the **number operator**, which is treated as a new dynamical observable.

Consider the Hilbert space spanned by the eigenvectors of the number operator. The properties concerning the eigenvalues as well as the eigenvectors are summarized in the following propositions.

Proposition 4.

If ν and $|\nu\rangle$ are respectively the eigenvalue and eigenvector of number operator N, then ν is positive definite.

For the eigenvalue equation of operator N,

$$N|\nu\rangle = \nu|\nu\rangle, \tag{2.60}$$

where the eigenvector is labeled by the eigenvalue and denoted it by $|\nu\rangle$. Let us define the vector $|\alpha\rangle = a|\nu\rangle$. The positive definite of the norm of the vector $|\alpha\rangle$ requires that

$$\langle\alpha|\alpha\rangle = \langle\nu|a^\dagger a|\nu\rangle = \nu\langle\nu|\nu\rangle \geqslant 0.$$

Therefore $\nu \geqslant 0$, namely ν is positive definite.

> **Proposition 5.**
>
> *The state with zero eigenvalue i.e.* $|0\rangle$, *then* $|0\rangle$ *will be annihilated into the null vector by* $\boldsymbol{a}$, *the annihilation operator, namely* $\boldsymbol{a}|0\rangle = 0$.

Let $|\alpha\rangle = \boldsymbol{a}|0\rangle$, and the norm square of $|\alpha\rangle$ reads as

$$\langle\alpha|\alpha\rangle = \langle 0|\boldsymbol{a}^\dagger\boldsymbol{a}|0\rangle = \langle 0|\boldsymbol{N}|0\rangle = 0\langle 0|0\rangle = 0, \qquad (2.61)$$

which implies that $\boldsymbol{a}|0\rangle$ is a null vector.

> **Proposition 6.**
>
> $\boldsymbol{a}^\dagger|\nu\rangle$ *is an eigenvector of* $\boldsymbol{N}$ *with eigenvalue* $\nu + 1$.

This can be proved with the commutation relation

$$[\boldsymbol{N}, \boldsymbol{a}^\dagger] = \boldsymbol{a}^\dagger, \qquad \text{or} \quad \boldsymbol{N}\boldsymbol{a}^\dagger = \boldsymbol{a}^\dagger(\boldsymbol{N} + \mathbf{I}). \qquad (2.62)$$

For $|\nu\rangle \neq 0$ and letting $|\alpha\rangle = \boldsymbol{a}^\dagger|\nu\rangle$, then

$$\boldsymbol{N}|\alpha\rangle = \boldsymbol{N}\boldsymbol{a}^\dagger|\nu\rangle = \boldsymbol{a}^\dagger(\boldsymbol{N} + \mathbf{I})|\nu\rangle = \boldsymbol{a}^\dagger(\nu + 1)|\nu\rangle = (\nu + 1)|\alpha\rangle. \quad (2.63)$$

> **Proposition 7.**
>
> *For* $|\nu\rangle \neq 0$, *then* $\boldsymbol{a}|\nu\rangle$ *is the eigenvector of* $\boldsymbol{N}$ *with eigenvalue* $\nu - 1$.

Since $[\boldsymbol{N}, \boldsymbol{a}] = -\boldsymbol{a}$, or $\boldsymbol{N}\boldsymbol{a} = \boldsymbol{a}(\boldsymbol{N} - \mathbf{I})$, and letting $|\alpha\rangle = \boldsymbol{a}|\nu\rangle$, then

$$\boldsymbol{N}|\alpha\rangle = \boldsymbol{N}\boldsymbol{a}|\nu\rangle = \boldsymbol{a}(\boldsymbol{N} - \mathbf{I})|\nu\rangle = \boldsymbol{a}(\nu - 1)|\nu\rangle = (\nu - 1)|\alpha\rangle. \qquad (2.64)$$

Proposition 8.

The eigenvalues of the number operator N take only non-negative integers, i.e. $\nu = n = 0, 1, 2, \ldots$.

The proof can be facilitated with Proposition 7. If $\nu \neq$ integer, then vectors $a|\nu\rangle, a^2|\nu\rangle, a^3|\nu\rangle, \ldots, a^s|\nu\rangle$, are all eigenvectors of N respectively with the eigenvalues $\nu - 1, \nu - 2, \ldots, \nu - s$. Let $|\alpha\rangle = a|\nu - s\rangle$, we calculate the square norm of $|\alpha\rangle$, i.e.

$$\langle\alpha|\alpha\rangle = (\nu - s)\langle\nu - s|\nu - s\rangle.$$

This does not guarantee that $\langle\alpha|\alpha\rangle$ is positive definite if s is chosen larger than ν. Positive definite of the norms of all eigenvectors generated by means of applying a^s on $|\nu\rangle$ can only be achieved if ν takes non-negative integers, namely $\nu = n = 0, 1, 2, \ldots$.

Therefore the eigenvectors of the number operator N, as well as that of the new Hamiltonian are written as

$$|0\rangle, |1\rangle, |2\rangle, \ldots, |n\rangle, \ldots$$

with the normalization $\langle n|n'\rangle = \delta_{nn'}$.

The eigenvector with zero eigenvalue, i.e. $|0\rangle$ is called the ground state of N or $\mathcal{H}$. The eigenvector $|1\rangle$ is called the one particle state for operator N, or the first excited state of the Hamiltonian operator $\mathcal{H}$ and so forth. This is the reason that $a^\dagger$ is called the creation operator or the promotion operator, while a is called the annihilation operator or the demotion operator. $a^\dagger$ acts on state $|n\rangle$ to produce the state $c|n+1\rangle$, i.e.

$$a^\dagger|n\rangle = c|n+1\rangle,$$

so that the action of $a^\dagger$ will increase the n-particle system into the $(n+1)$-particle system or will promote the n-th excited state into $(n+1)$-th excited state. The constant c is determined to be $\sqrt{n+1}$ when all eigenstates are orthonormalized, i.e.

$$a^\dagger|n\rangle = \sqrt{n+1}|n+1\rangle. \tag{2.65}$$

Similarly operator a acts upon state $|n\rangle$ and it becomes the state $c'|n-1\rangle$, i.e.

$$a|n\rangle = c'|n-1\rangle.$$

The action of a decreases the n-particle system into the $(n-1)$-particle system or demotes the n-th excited state into the $(n-1)$-th excited state. Again the constant c' can be evaluated to be $\sqrt{n}$, i.e.

$$a|n\rangle = \sqrt{n}|n-1\rangle. \tag{2.66}$$

With the help of the creation operator, one is able to construct the orthonormal basis of the Hilbert space spanned by the eigenvectors of the number operator N. We start from the ground state or zero particle states and build the higher excited states or many particle state by acting the creation operator upon it successively. Then we have

$$|0\rangle,$$

$$|1\rangle = a^\dagger|0\rangle,$$

$$|2\rangle = \frac{1}{\sqrt{2!}}(a^\dagger)^2|0\rangle,$$

$$\vdots$$

$$|n\rangle = \frac{1}{\sqrt{n!}}(a^\dagger)^n|0\rangle. \tag{2.67}$$

Conversely, we are also able to construct the lower excited states by acting the annihilation operator upon the n-particle states successively, i.e.

$$|0\rangle = \boldsymbol{a}|1\rangle,$$

$$|1\rangle = \frac{1}{\sqrt{2}}\boldsymbol{a}|2\rangle,$$

$$|2\rangle = \frac{1}{\sqrt{3}}\boldsymbol{a}|3\rangle,$$

$$\vdots$$

$$|n\rangle = \frac{1}{\sqrt{n+1}}\boldsymbol{a}|n+1\rangle. \tag{2.68}$$

2.4 N-representation and the Rodrigues formula of Hermite polynomials

It is interesting to investigate how the eigenstate $|n\rangle$ is to be realized in the q-representation. Let us consider the eigenvalue equation of the operator $\boldsymbol{\xi}$, i.e.

$$\boldsymbol{\xi}|\xi\rangle = \xi|\xi\rangle. \tag{2.69}$$

One can also easily verify that

$$\langle\xi|\boldsymbol{p}_\xi|\xi'\rangle = -i\frac{d}{d\xi}\delta(\xi - \xi'),$$

with the normalization $\langle\xi|\xi'\rangle = \delta(\xi - \xi')$.

The inner product of $|\xi\rangle$ and $|n\rangle$ defines the q-representation of the n-particle state, namely the n-particle wave function or the wave function of the n-th excited state, which reads as

$$\psi_n(\xi) = \langle\xi|n\rangle.$$

Since $a^\dagger = \dfrac{1}{\sqrt{2}}(\boldsymbol{\xi} - i\boldsymbol{p}_\xi)$, the wave function $\psi_n(\xi)$ becomes

$$\psi_n(\xi) = \langle \xi | n \rangle = \frac{1}{\sqrt{n!}} \langle \xi | (a^\dagger)^n | 0 \rangle = \frac{1}{\sqrt{n!}} \left(\frac{1}{\sqrt{2}} \right)^n \langle \xi | (\boldsymbol{\xi} - i\boldsymbol{p}_\xi)^n | 0 \rangle.$$

By Proposition 2, we have

$$\langle \xi | (\boldsymbol{\xi} - i\boldsymbol{p}_\xi)^n | 0 \rangle = \left(\xi - \frac{d}{d\xi} \right)^n \langle \xi | 0 \rangle = \left(\xi - \frac{d}{d\xi} \right)^n \psi_0(\xi).$$

To obtain the 0-particle wave function $\psi_0(\xi)$, we take the q-representation of $\boldsymbol{a}|0\rangle = \boldsymbol{0}$, i.e.

$$\langle \xi | \boldsymbol{a} | 0 \rangle = 0,$$

namely that

$$\langle \xi | \frac{1}{\sqrt{2}} (\boldsymbol{\xi} + i\boldsymbol{p}_\xi) | 0 \rangle = 0,$$

or

$$\left(\xi + \frac{d}{d\xi} \right) \psi_0(\xi) = 0. \tag{2.70}$$

Therefore we can solve the equation and obtain the normalized wave function as

$$\psi_0(\xi) = \frac{1}{\sqrt[4]{\pi}} e^{-\frac{1}{2}\xi^2}, \tag{2.71}$$

and the wave function $\psi_n(\xi)$ becomes

$$\psi_n(\xi) = \frac{1}{\sqrt{n!}} \left(\frac{1}{\sqrt{2}} \right)^n \frac{1}{\sqrt[4]{\pi}} \left(\xi - \frac{d}{d\xi} \right)^n e^{-\frac{1}{2}\xi^2}. \tag{2.72}$$

Applying the following identity

$$\xi - \frac{d}{d\xi} \equiv (-1)e^{\frac{1}{2}\xi^2}\frac{d}{d\xi}e^{-\frac{1}{2}\xi^2},$$

we conclude that

$$\left(\xi - \frac{d}{d\xi}\right)^n = (-1)^n e^{\frac{1}{2}\xi^2}\left(\frac{d}{d\xi}\right)^n e^{-\frac{1}{2}\xi^2}.$$

Then

$$\psi_n(\xi) = \frac{(-1)^n}{\sqrt{2^n n!\sqrt{\pi}}} e^{\frac{1}{2}\xi^2}\left(\frac{d}{d\xi}\right)^n e^{-\xi^2} = \frac{e^{-\frac{1}{2}\xi^2}}{\sqrt{2^n n!\sqrt{\pi}}} H_n(\xi), \qquad (2.73)$$

and the Rodrigues formula of Hermite polynomials is identified as

$$H_n(\xi) = (-1)^n e^{\xi^2}\left(\frac{d}{d\xi}\right)^n e^{-\xi^2}. \qquad (2.74)$$

2.5 Two dimensional harmonic oscillation and direct product of vector spaces

Let us consider the system of a two-dimensional isotropic harmonic oscillator whose Hamiltonian takes the following expression

$$\mathbf{H} = \frac{1}{2m}(\mathbf{P}_x^2 + \mathbf{P}_y^2) + \frac{m}{2}\omega(\mathbf{X}^2 + \mathbf{Y}^2). \qquad (2.75)$$

Similar to the case of one-dimensional harmonic oscillation, we define the creation operators and the annihilation operators as follows:

$$a^\dagger = \frac{1}{\sqrt{2}}(\boldsymbol{\xi} - i\boldsymbol{p}_\xi), \tag{2.76a}$$

$$a = \frac{1}{\sqrt{2}}(\boldsymbol{\xi} + i\boldsymbol{p}_\xi), \tag{2.76b}$$

$$b^\dagger = \frac{1}{\sqrt{2}}(\boldsymbol{\eta} - i\boldsymbol{p}_\eta), \tag{2.76c}$$

$$b = \frac{1}{\sqrt{2}}(\boldsymbol{\eta} + i\boldsymbol{p}_\eta), \tag{2.76d}$$

where ξ, η, p_ξ and p_η are defined as

$$\boldsymbol{\xi} = \sqrt{\frac{m\omega}{\hbar}}\mathbf{X}, \quad \boldsymbol{p}_\xi = \frac{1}{\sqrt{m\hbar\omega}}\mathbf{P}_x, \tag{2.77a}$$

$$\boldsymbol{\eta} = \sqrt{\frac{m\omega}{\hbar}}\mathbf{Y}, \quad \boldsymbol{p}_\eta = \frac{1}{\sqrt{m\hbar\omega}}\mathbf{P}_y. \tag{2.77b}$$

The fundamental commutation relations can be simplified as

$$[\boldsymbol{\xi}, \boldsymbol{\eta}] = [\boldsymbol{\xi}, \boldsymbol{p}_\eta] = \mathbf{O},$$
$$[\boldsymbol{p}_\xi, \boldsymbol{\eta}] = [\boldsymbol{p}_\xi, \boldsymbol{p}_\eta] = \mathbf{O},$$
$$[\boldsymbol{p}_\xi, \boldsymbol{\xi}] = [\boldsymbol{p}_\eta, \boldsymbol{\eta}] = -i\mathbf{I},$$

which lead to the following commutation relations among the annihilation operators and the creation operators, namely

$$[a, b] = [a, b^\dagger] = [a^\dagger, b] = [a^\dagger, b^\dagger] = \mathbf{O}, \tag{2.78}$$

$$[a, a^\dagger] = [b, b^\dagger] = \mathbf{I}. \tag{2.79}$$

The Hamiltonian can therefore be written as

$$\mathbf{H} - \left(a^\dagger a + \frac{1}{2}\mathbf{I}\right)\hbar\omega + \left(b^\dagger b + \frac{1}{2}\mathbf{I}\right)\hbar\omega,$$

or $\mathbf{H} = \mathbf{H}_a + \mathbf{H}_b$ if we put

$$\mathbf{H}_a = \left(a^\dagger a + \frac{1}{2}\mathbf{I}\right)\hbar\omega = \left(\mathbf{N}_a + \frac{1}{2}\mathbf{I}\right)\hbar\omega,$$

$$\mathbf{H}_b = \left(b^\dagger b + \frac{1}{2}\mathbf{I}\right)\hbar\omega = \left(\mathbf{N}_b + \frac{1}{2}\mathbf{I}\right)\hbar\omega.$$

To solve the Schrödinger equation of the two dimensional harmonic oscillation, we shall make use of Proposition 8 in Section 1.6.2. Evaluating the commutator of the operator $\mathbf{N}_a$ and the operator $\mathbf{N}_b$, we find that only the first and the second fundamental commutation relations are required to show that $[\mathbf{N}_a, \mathbf{N}_b] = \mathbf{O}$. Therefore we recognize that $\mathbf{N}_a$ and $\mathbf{N}_b$ are, in fact, the intrinsic compatible observables. The eigenvectors $|\psi\rangle$ in the following Schrödinger equation

$$\left\{\left(\mathbf{N}_a + \frac{1}{2}\mathbf{I}\right)\hbar\omega + \left(\mathbf{N}_b + \frac{1}{2}\mathbf{I}\right)\hbar\omega\right\}|\psi\rangle = E|\psi\rangle, \qquad (2.80)$$

can be expressed as the direct product of the two eigenvectors in one-dimensional case, i.e.

$$|\psi\rangle = |n_a\rangle \otimes |n_b\rangle \equiv |n_a; n_b\rangle, \qquad (2.81)$$

with the corresponding eigenvalues given as

$$E = (n_a + n_b + 1)\hbar\omega = (n + 1)\hbar\omega. \qquad (2.82)$$

The energy of the two-dimensional harmonic oscillator is also quantized to be $(n + 1)\hbar\omega$, with degeneracy if $n_a + n_b = n > 0$.

We are now in the position to discuss the algebra of the direct vector space. A direct product space of the Hilbert spaces $\mathcal{H}_a$ and $\mathcal{H}_b$ is also a Hilbert space, denoted as follows

$$\mathcal{H} = \mathcal{H}_a \otimes \mathcal{H}_b. \qquad (2.83)$$

Let $\mathcal{H}_a$ be the Hilbert space spanned by $|a_i\rangle$, the eigenvectors of the dynamical operator $\mathbf{A}$, then the basis of $\mathcal{H}_a$ takes the following set of vectors,

$$|a_1\rangle, |a_2\rangle, \ldots |a_i\rangle, \ldots$$

Similarly, if the Hilbert space $\mathcal{H}_b$ is spanned by another set of basis, denoted by

$$|b_1\rangle, |b_2\rangle, \ldots |b_j\rangle, \ldots$$

which are the eigenvectors of another dynamical operator $\mathbf{B}$. The product space $\mathcal{H} = \mathcal{H}_a \otimes \mathcal{H}_b$ is then spanned by the set of direct product of a base vector $|a_i\rangle$ in $\mathcal{H}_a$ and another base vector $|b_j\rangle$ in $\mathcal{H}_b$, denoted by

$$|a_i\rangle \otimes |b_j\rangle \equiv |a_i; b_j\rangle. \tag{2.84}$$

This very **direct product vector** $|a_i; b_j\rangle$ is also the eigenvector of any linear operator in $\mathcal{H}$ if the operator is constructed in the form of a function of two operators $\mathbf{A}$ and $\mathbf{B}$, denoted by $\mathbf{F}(\mathbf{A}, \mathbf{B})$, and the eigenvalue equation for the operator $\mathbf{F}(\mathbf{A}, \mathbf{B})$ reads as follows

$$\mathbf{F}(\mathbf{A}, \mathbf{B})|a_i; b_j\rangle = F(a_i, b_j)|a_i; b_j\rangle, \tag{2.85}$$

where $F(a_i, b_j)$ is the eigenvalue of $\mathbf{F}(\mathbf{A}, \mathbf{B})$.

The dimension of the product space $\mathcal{H}$ equals to the product of the dimensions of the two individual spaces $\mathcal{H}_a$ and $\mathcal{H}_b$, namely $d = d_a d_b$ with d_a, d_b and d standing for the dimensions of $\mathcal{H}_a, \mathcal{H}_b$ and $\mathcal{H}$ respectively.

The linear operator in $\mathcal{H}$ also takes the form of direct product of two linear operators, denoted by $\mathbf{R}$ and $\mathbf{S}$, where $\mathbf{R}$ is a linear operator in the space $\mathcal{H}_a$ and $\mathbf{S}$ is a linear operator in the space $\mathcal{H}_b$. The direct product operator is commonly denoted by $\mathbf{R} \otimes \mathbf{S}$, whose algebraic operational rule reads as

$$\mathbf{R} \otimes \mathbf{S}|a_i; b_j\rangle = \mathbf{R}|a_i\rangle \otimes \mathbf{S}|b_j\rangle. \tag{2.86}$$

We summarize the properties of the direct product operators as follows:

$$\mathbf{R} \otimes \mathbf{O} = \mathbf{O} \otimes \mathbf{S} = \mathbf{O}, \tag{2.87a}$$

$$\mathbf{I} \otimes \mathbf{I} = \mathbf{I}, \mathbf{R} \otimes \mathbf{I} = \mathbf{R}, \mathbf{I} \otimes \mathbf{S} = \mathbf{S}, \tag{2.87b}$$

$$\mathbf{R} \otimes (\mathbf{S}_1 + \mathbf{S}_2) = \mathbf{R} \otimes \mathbf{S}_1 + \mathbf{R} \otimes \mathbf{S}_2, \tag{2.87c}$$

$$(\mathbf{R}_1 + \mathbf{R}_2) \otimes \mathbf{S} = \mathbf{R}_1 \otimes \mathbf{S} + \mathbf{R}_2 \otimes \mathbf{S}, \tag{2.87d}$$

$$\alpha \mathbf{R} \otimes \beta \mathbf{S} = \alpha\beta \mathbf{R} \otimes \mathbf{S}, \tag{2.87e}$$

$$(\mathbf{R} \otimes \mathbf{S})^{-1} = \mathbf{R}^{-1} \otimes \mathbf{S}^{-1}, \tag{2.87f}$$

$$\mathbf{R}_1 \mathbf{R}_2 \otimes \mathbf{S}_1 \mathbf{S}_2 = (\mathbf{R}_1 \otimes \mathbf{S}_1)(\mathbf{R}_2 \otimes \mathbf{S}_2), \tag{2.87g}$$

where α and β are scalars.

To demonstrate the difference between the compatible observables and the intrinsic compatible ones, we take the motion of a charged particle in the uniform magnetic field as an example. Consider a uniform magnetic field of H_0 in z-axis, the vector potential $\vec{A}(x, y)$ can be expressed as

$$\vec{A}(x, y) = \frac{1}{2} H_0(-y\hat{i} + x\hat{j}).$$

The principle of minimal interaction allows us to cast the Hamiltonian of a particle, with the mass m and the charge q moving in the uniform field H_0, into the following form

$$\mathbf{H} = \frac{1}{2m} \left(\mathbf{P} - \frac{q}{c}\mathbf{A} \right)^2$$

$$= \frac{1}{2m} \left\{ \left(\mathbf{P}_x + \frac{1}{2}\frac{q}{c}H_0\mathbf{Y} \right)^2 + \left(\mathbf{P}_y - \frac{1}{2}\frac{q}{c}H_0\mathbf{X} \right)^2 \right\}. \tag{2.88}$$

The sum of the two square terms in the Hamiltonian of last equation ensures that the total energy of the system is always positive definite. Namely when we solve the following Schrödinger equation

$$\mathbf{H}|\psi\rangle = E|\psi\rangle, \tag{2.89}$$

we shall obtain the positive definite eigenvalue E.

By expanding the quadratic terms of the above Hamiltonian, we obtain the following expression, i.e.

$$\mathbf{H} = \frac{1}{2m}\mathbf{P}_x^2 + \frac{m}{2}\left(\frac{\omega_c}{2}\right)^2\mathbf{X}^2 + \frac{1}{2m}\mathbf{P}_y^2 + \frac{m}{2}\left(\frac{\omega_c}{2}\right)^2\mathbf{Y}^2 - \frac{1}{2}\omega_c\mathbf{L}_z, \quad (2.90)$$

where ω_c is the **cyclotron frequency** and $\mathbf{L}_z$ stands for the 3rd component of the **angular momentum operator** which can be calculated explicitly as follows:

$$\omega_c = \frac{qH_0}{mc}, \quad \mathbf{L}_z = \mathbf{XP}_y - \mathbf{YP}_x.$$

One recognizes immediately that combining the first two terms of Eq. (2.90) is , in fact, the Hamiltonian of a two dimensional harmonic oscillator with the force constant $k = \frac{m}{2}\left(\frac{\omega_c}{2}\right)^2$, i.e.

$$\mathbf{H}_{h.o.}^{(2)} = \frac{1}{2m}(\mathbf{P}_x^2 + \mathbf{P}_y^2) + \frac{m}{2}\left(\frac{\omega_c}{2}\right)^2(\mathbf{X}^2 + \mathbf{Y}^2). \quad (2.91)$$

It can be easily verified that $\mathbf{H}_{h.o.}^{(2)}$ and $\mathbf{L}_z$ commute with each other i.e.

$$[\mathbf{H}_{h.o.}^{(2)}, \mathbf{L}_z] = \mathbf{O}. \quad (2.92)$$

We will run into a dilemma if we apply the method of Proposition 8 to calculate the eigenvalues of the following Hamiltonian

$$\mathbf{H} = \mathbf{H}_{h.o.}^{(2)} - \frac{1}{2}\omega_c\mathbf{L}_z, \quad (2.93)$$

which is the sum of two compatible observables. The eigenvalue equations for these two observables read as

$$\mathbf{H}_{h.o.}^{(2)} |n_1; n_2\rangle = \frac{1}{2}(n_1 + n_2 + 1)\hbar\omega_c |n_1; n_2\rangle, \qquad (2.94)$$

$$\mathbf{L}_z |m\rangle = m\hbar |m\rangle, \qquad (2.95)$$

where m takes the integers, while n_1 and n_2 take only the positive ones. Of course we have made use of the result that the angular momentum $\mathbf{L}_z$ is quantized as it is expressed in Eq. (2.95), which we shall investigate in the next chapter.

Should we try to evaluate the total energy of the system by applying $\mathbf{H} = \mathbf{H}_{h.o.}^{(2)} - \omega_c \mathbf{L}_z / 2$ to act upon the direct product vector $|n_1; n_2\rangle \otimes |m\rangle$, we would end up in the following eigenvalue equation

$$\left(\mathbf{H}_{h.o.}^{(2)} - \frac{1}{2}\omega_c \mathbf{L}_z \right) |n_1; n_2\rangle \otimes |m\rangle = E |n_1; n_2\rangle \otimes |m\rangle, \qquad (2.96)$$

with $E = (n_1 + n_2 + 1)\hbar\omega_c/2 - m\hbar\omega_c/2$, which is no longer a positive definite quantity if $m > (n_1 + n_2 + 1)$. Namely the total energy of the system E obtained in this way does not guarantee to be positive definite, which contradicts our previous assertion.

The contradiction arises from the fact that the observable $H_{h.o.}^{(2)}$ and the observable $\mathbf{L}_z$ are not intrinsic compatible even though they commute with each other. One needs, in fact, all the three fundamental commutation relations in order to prove that the commutator of $\mathbf{H}_{h.o.}^{(2)}$ and $\mathbf{L}_z$ vanishes, i.e.

$$[\mathbf{H}_{h.o.}^{(2)}, \mathbf{L}_z] = \mathbf{O}. \qquad (2.97)$$

The dilemma can be resolved by introducing another set of linear operators in terms of the original four dynamical observables with the following relations:

$$c = \frac{1}{\sqrt{2}} \left\{ \sqrt{\frac{m\omega_c}{2\hbar}}(\mathbf{X} + i\mathbf{Y}) + i\sqrt{\frac{2}{m\hbar\omega_c}}(\mathbf{P}_x - i\mathbf{P}_y) \right\}, \qquad (2.98a)$$

$$c^\dagger = \frac{1}{\sqrt{2}} \left\{ \sqrt{\frac{m\omega_c}{2\hbar}}(\mathbf{X} - i\mathbf{Y}) - i\sqrt{\frac{2}{m\hbar\omega_c}}(\mathbf{P}_x + i\mathbf{P}_y) \right\}, \qquad (2.98b)$$

$$d = \frac{1}{\sqrt{2}} \left\{ \sqrt{\frac{m\omega_c}{2\hbar}}(\mathbf{X} - i\mathbf{Y}) + i\sqrt{\frac{2}{m\hbar\omega_c}}(\mathbf{P}_x + i\mathbf{P}_y) \right\}, \qquad (2.98c)$$

$$d^\dagger = \frac{1}{\sqrt{2}} \left\{ \sqrt{\frac{m\omega_c}{2\hbar}}(\mathbf{X} + i\mathbf{Y}) - i\sqrt{\frac{2}{m\hbar\omega_c}}(\mathbf{P}_x - i\mathbf{P}_y) \right\}. \qquad (2.98d)$$

It can be verified that the commutation relations among c, $c^\dagger$, d and $d^\dagger$ become as follows:

$$[c, d] = [c, d^\dagger] = \mathbf{O}, \qquad (2.99a)$$

$$[c^\dagger, d] = [c^\dagger, d^\dagger] = \mathbf{O}, \qquad (2.99b)$$

$$[c^\dagger, c] = [d^\dagger, d] = \mathbf{I}. \qquad (2.99c)$$

If we construct two Hermitian number operators defined as $\mathbf{N}_c = c^\dagger c$ and $\mathbf{N}_d = d^\dagger d$. It is obvious that $\mathbf{N}_c$ and $\mathbf{N}_d$ become the intrinsic compatible observables because that $[\mathbf{N}_c, \mathbf{N}_d] = \mathbf{O}$ is derived only based upon the first two sets of the commutation relations, i.e. Eqs. (2.99a) and (2.99b). Therefore we can construct the direct product vector $|n_c; n_d\rangle = |n_c\rangle \otimes |n_d\rangle$, where $|n_c\rangle$ and $|n_d\rangle$ stand for the eigenvectors of the operators $\mathbf{N}_c$ and $\mathbf{N}_d$ respectively. Using the property that $|n_c; n_d\rangle$ is automatically the eigenvector of the of the operator $\mathbf{F}(\mathbf{N}_c, \mathbf{N}_d)$ with the eigenvalue $F(n_c, n_d)$, we are able to calculate the eigenvalue E of the Hamiltonian operator $\mathbf{H}$ in the following equation

$$\left(\mathbf{H}_{h.o.}^{(2)} - \frac{1}{2}\omega_c \mathbf{L}_z \right) |n_c; n_d\rangle = E|n_c; n_d\rangle. \qquad (2.100)$$

After some algebra, the two terms $\mathbf{H}_{h.o.}^{(2)}$ and $\mathbf{L}_z$ in the Hamiltonian above can be expressed in terms of the new dynamical operators $\mathbf{N}_c$ and $\mathbf{N}_d$ as follows:

$$\mathbf{H}_{h.o.}^{(2)} = \frac{1}{2}(\mathbf{N}_c + \mathbf{N}_d + 1)\hbar\omega_c, \tag{2.101}$$

$$-\frac{1}{2}\omega_c\mathbf{L}_z = \frac{1}{2}(\mathbf{N}_c - \mathbf{N}_d)\hbar\omega_c, \tag{2.102}$$

which allow us to take $\mathbf{F}(\mathbf{N}_c, \mathbf{N}_d)$ as follow

$$\mathbf{F}(\mathbf{N}_c, \mathbf{N}_d) = \mathbf{H}_{h.o.}^{(2)} - \frac{1}{2}\omega_c\mathbf{L}_z = \left(\mathbf{N}_c + \frac{1}{2}\right)\hbar\omega_c, \tag{2.103}$$

and the eigenvalue E is evaluated as

$$E = F(n_c, n_d) = \left(n_c + \frac{1}{2}\right)\hbar\omega_c, \tag{2.104}$$

which is positive definite as we have expected at the beginning of our discussion on the system of a charged particle moving in the uniform magnetic field.

2.6 Elastic medium and the quantization of scalar field

Consider the linear chain of N identical particles of mass m, with the springs of force constant α attached to both sides of the nearest mass points. Each mass point is also attached with another spring of force constant β to the equilibrium positions which are of equal spacing along the chain, as shown in Figure 2.1. The Hamiltonian of the system reads as

$$H = \frac{1}{2m}\sum_{r=1}^{N}p_r^2 + \frac{\alpha}{2}\sum_{r=1}^{N}(q_r - q_{r+1})^2 + \frac{\beta}{2}\sum_{r=1}^{N}q_rq_r, \tag{2.105}$$

where q_r is the displacement of the r-th mass point away from its equilibrium position, and p_r is the corresponding canonical conjugate momentum. The commutation relations are expressed as

$$[\boldsymbol{q}_r, \boldsymbol{q}_{r'}] = [\boldsymbol{p}_r, \boldsymbol{p}_{r'}] = \mathbf{O}, \qquad [\boldsymbol{q}_r, \boldsymbol{p}_{r'}] = i\hbar \delta_{rr'} \mathbf{I}.$$

The Hamiltonian can be simplified tremendously if we introduce the **normal coordinates** denoted by $\boldsymbol{Q}_s$ through the following Fourier expansion

$$\boldsymbol{q}_r = \sqrt{\frac{\hbar}{Nm}} \sum_s e^{\frac{2\pi i}{N} rs} \boldsymbol{Q}_s. \tag{2.106}$$

Here we assume the total number of mass points to be N, and

$$s = \begin{cases} \text{takes from } s = -\frac{1}{2}(N-1) \text{ to } s = \frac{1}{2}(N-1), \\ \text{they are integers if } N \text{ is odd, and half odd integer if } N \text{ is even.} \end{cases}$$

The condition that $\boldsymbol{q}_r$ is real will lead us to take $\boldsymbol{Q}_{-s} = \boldsymbol{Q}_s^\dagger$. Before we try to decouple the mixing terms in the Hamiltonian, we establish the following formula:

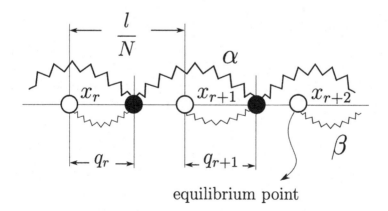

equilibrium point

Figure 2.1: Linear chain of identical particles

Proposition 9.

For r = integers, and $s - s'$ = integers, then

$$\frac{1}{N}\sum_{r=1}^{N} e^{\frac{2\pi i}{N}r(s-s')} = \delta_{ss'}.$$

Denoting $z = e^{\frac{2\pi i}{N}(s-s')}$, we have

$$F(z) = \frac{1}{N}\sum_r z^r = \frac{1}{N}(z + z^2 + \ldots + z^N) = \frac{z(1-z^N)}{N(1-z)}, \qquad (2.107)$$

which implies that for the case $s - s' = 0$,

$$F(z) = F(1) = 1.$$

For the case that $s - s'$ = non-zero integers, say $s - s' = m$ then $z = \exp(2\pi i N/m)$, which is one of the roots for the equation $z^N - 1 = 0$. Therefore

$$F(z) = \frac{1}{N}(1 + z + \ldots + z^{N-1}) = 0,$$

and one reaches the conclusion

$$F(z) = \delta_{ss'}. \qquad (2.108)$$

This proposition allows us to invert q_r into Q_s. We leave it as an exercise to invert Eq. (2.106) and obtain that

$$Q_s = \sqrt{\frac{m}{N\hbar}}\sum_r e^{-\frac{2\pi i}{N}rs}q_r. \qquad (2.109)$$

Similarly the Fourier expansion for the momentum is expressed as

$$p_r = \sqrt{\frac{m\hbar}{N}} \sum_s e^{\frac{2\pi i}{N} rs} P_s, \qquad (2.110)$$

with

$$P_{-s} = P_s^\dagger,$$

and

$$P_s = \frac{1}{\sqrt{mN\hbar}} \sum_r e^{-\frac{2\pi i}{N} rs} p_r. \qquad (2.111)$$

It also leaves you as an exercise to show that each term in the Hamiltonian operator can be calculated as follows,

$$\frac{1}{2m} \sum_r p_r^2 = \frac{\hbar}{2} \sum_s P_s P_s^\dagger, \qquad (2.112)$$

$$\frac{\alpha}{2} \sum_r (q_r - q_{r+1})^2 = \frac{\alpha\hbar}{2m} \sum_s 4\sin^2\left(\frac{2\pi s}{N}\right) Q_s Q_s^\dagger, \qquad (2.113)$$

$$\frac{\beta}{2} \sum_r q_r q_r = \frac{\beta\hbar}{2m} \sum_s Q_s Q_s^\dagger. \qquad (2.114)$$

Therefore the Hamiltonian takes the following expression

$$\mathbf{H} = \frac{\hbar}{2} \sum [P_s P_s^\dagger + \omega^2(s) Q_s Q_s^\dagger], \qquad (2.115)$$

where $\omega^2(s)$ is defined as

$$\omega^2(s) = \frac{4\alpha}{m} \sin^2\left(\frac{2\pi s}{N}\right) + \frac{\beta}{m}, \qquad (2.116)$$

and $\omega(s)$ is an even function in s, i.e.

$$\omega(-s) = \omega(s).$$

It can be verified that the commutation relations for the operators expressed in terms of normal coordinate and momentum become

$$[\boldsymbol{Q}_s, \boldsymbol{Q}_{s'}] = [\boldsymbol{P}_s, \boldsymbol{P}_{s'}] = \mathbf{O}, \qquad [\boldsymbol{Q}_s, \boldsymbol{P}_{s'}] = i\delta_{ss'}\mathbf{I}. \qquad (2.117)$$

Let us investigate the system in the Heisenberg picture in which the operators are regarded as time dependent. The equations of motion for $\boldsymbol{Q}_s$, $\boldsymbol{Q}_s^\dagger$, $\boldsymbol{P}_s$ and $\boldsymbol{P}_s^\dagger$ take the identical expressions as the canonical equations of motion in classical mechanics, namely,

$$\frac{\partial \mathbf{H}}{\partial \boldsymbol{P}_s^\dagger} = \dot{\boldsymbol{Q}}_s^\dagger, \quad \text{or} \quad \boldsymbol{P}_s = \dot{\boldsymbol{Q}}_s^\dagger,$$

$$\frac{\partial \mathbf{H}}{\partial \boldsymbol{Q}_s^\dagger} = -\dot{\boldsymbol{P}}_s^\dagger, \quad \text{or} \quad \omega^2(s)\boldsymbol{Q}_s = -\dot{\boldsymbol{P}}_s^\dagger.$$

Hence we have

$$\ddot{\boldsymbol{Q}}_s + \omega^2(s)\boldsymbol{Q}_s = 0, \qquad (2.118)$$

and the solution is cast into

$$\boldsymbol{Q}_s = \frac{1}{\sqrt{2\omega(s)}}\left(\boldsymbol{a}_s e^{-i\omega(s)t} + \boldsymbol{a}_{-s}^\dagger e^{i\omega(s)t}\right), \qquad (2.119)$$

where $\boldsymbol{a}_s/\sqrt{2\omega(s)}$ and $\boldsymbol{a}_{-s}^\dagger/\sqrt{2\omega(s)}$, playing the roles of the constants of integration, are chosen such as to meet the condition $\boldsymbol{Q}_s = \boldsymbol{Q}_{-s}^\dagger$.

Similarly we have the solution for $\boldsymbol{P}_s$ as

$$\boldsymbol{P}_s = i\sqrt{\frac{\omega(s)}{2}}\left(-\boldsymbol{a}_s e^{-i\omega(s)t} + \boldsymbol{a}_{-s}^\dagger e^{i\omega(s)t}\right). \qquad (2.120)$$

Let us define $b_s = a_s e^{-i\omega(s)t}$ and $b_{-s}^\dagger = a_{-s}^\dagger e^{i\omega(s)t}$. Then the normal coordinate and the normal momentum become as follows

$$Q_s = \frac{1}{\sqrt{2\omega(s)}}(b_s + b_{-s}^\dagger),$$

$$P_s = i\sqrt{\frac{\omega(s)}{2}}(-b_s + b_{-s}^\dagger),$$

which enables us to write the commutation relations among the operators as

$$[b_s, b_{s'}] = [b_s^\dagger, b_{s'}^\dagger] = O, \qquad [b_s, b_{s'}^\dagger] = \delta_{ss'}I.$$

The Hamiltonian operator can then be expressed as

$$\mathbf{H} = \frac{\hbar}{2}\sum_s \omega(s)(b_s b_s^\dagger + b_s^\dagger b_s) = \hbar \sum_s \omega(s)\left(b_s^\dagger b_s + \frac{1}{2}\mathbf{I}\right)$$

$$= \hbar \sum_s \omega(s)\left(N_s + \frac{1}{2}\mathbf{I}\right), \qquad (2.121)$$

where $N_s = b_s^\dagger b_s$ is the boson number operator of s-th mode.

Let us investigate the system of an elastic medium in the limit of continuum. Take the length of the linear chain be l with N mass points equally spaced when they are at equilibrium. Then the position for the r-th particle is given as

$$x_r = r\frac{l}{N}. \qquad (2.122)$$

We also define the **wave number**

$$k_s = \frac{2\pi}{l}s, \qquad (2.123)$$

and rescale the displacement away from the r-th equilibrium position at the time t defined as

$$
\begin{aligned}
\boldsymbol{\Phi}(x_r, t) &= \sqrt{\frac{mN}{l}} \, \boldsymbol{q}_r = \sqrt{\frac{mN}{l}} \sqrt{\frac{\hbar}{mN}} \sum_s e^{\frac{2\pi i}{N} rs} \boldsymbol{Q}_s \\
&= \sqrt{\frac{\hbar}{l}} \sum_s e^{\frac{2\pi i}{N} rs} \frac{1}{\sqrt{2\omega(s)}} \left(\boldsymbol{a}_s e^{-i\omega(s)t} + \boldsymbol{a}_{-s}^\dagger e^{i\omega(s)t} \right) \\
&= \sqrt{\frac{\hbar}{l}} \sum_s \frac{1}{\sqrt{2\omega(s)}} \left(\boldsymbol{a}_s e^{i(k_s x_r - \omega(s)t)} + \boldsymbol{a}_s^\dagger e^{-i(k_s x_r - \omega(s)t)} \right),
\end{aligned}
$$

$$(2.124)$$

and the corresponding momentum as

$$
\boldsymbol{\Pi}(x_r, t) = \sqrt{\frac{\hbar}{l}} i \sum_s \sqrt{\frac{\omega(s)}{2}} \left(-\boldsymbol{a}_s e^{i(k_s x_r - \omega(s)t)} + \boldsymbol{a}_s^\dagger e^{-i(k_s x_r - \omega(s)t)} \right),
$$

$$(2.125)$$

where the last term in Eqs. (2.124) and (2.125) are obtained by making use of $k_s = -k_{-s}$ and $\omega(s) = \omega(-s)$.

The commutation relation between $\boldsymbol{\Phi}(x_r, t)$ and $\boldsymbol{\Pi}(x_{r'}, t)$ can be evaluated as

$$
[\boldsymbol{\Phi}(x_r, t), \boldsymbol{\Pi}(x_{r'}, t)] = \frac{\hbar}{l} N \delta_{rr'} \mathbf{I}.
$$

Let us take the limit as $N \to \infty$, the positions $x_r = rl/N$ become continuous, and the subscript r can be removed from x_r, namely that

$$
x = \lim_{N \to \infty} x_r = \lim_{N \to \infty} r \frac{l}{N}.
$$

$$(2.126)$$

We also simplify the wave number k_s into k, therefore

$$\Phi(x,t) = \lim_{N \to \infty} \Phi(x_r,t)$$

$$= \sqrt{\frac{\hbar}{l}} \sum_k \frac{1}{\sqrt{2\omega(k)}} \left(a_k e^{i(kx-\omega(k)t)} + a_k^\dagger e^{-i(kx-\omega(k)t)} \right), \quad (2.127)$$

$$\Pi(x,t) = \lim_{N \to \infty} \Pi(x_r,t)$$

$$= \sqrt{\frac{\hbar}{l}} \sum_k i\sqrt{\frac{\omega(k)}{2}} \left(-a_k e^{i(kx-\omega(k)t)} + a_k^\dagger e^{-i(kx-\omega(k)t)} \right),$$

$$(2.128)$$

and the commutation relation between $\Phi(x_r,t)$ and $\Pi(x_r,t)$ reduces to

$$[\Phi(x,t), \Pi(x',t)] = \lim_{N \to \infty} [\Phi(x_r,t), \Pi(x_{r'},t)]$$

$$= i\hbar \lim_{N \to \infty} \frac{N}{l} \delta_{rr'} \mathbf{I} = i\hbar \delta(x-x') \mathbf{I}. \quad (2.129)$$

It is on the quantum level that the continuous elastic medium is treated as a space being filled by the quantized scalar field or boson field.

The linear system can be generalized into the three-dimensional case. The field and its conjugate momentum take the expressions as

$$\Phi(\mathbf{r},t) = \sqrt{\frac{\hbar}{l}} \sum_{\mathbf{k}} \frac{1}{\sqrt{2\omega(\mathbf{k})}} \left(a_k e^{i(\mathbf{k}\mathbf{r}-\omega(\mathbf{k})t)} + a_k^\dagger e^{-i(\mathbf{k}\mathbf{r}-\omega(\mathbf{k})t)} \right), \quad (2.130)$$

$$\Pi(\mathbf{r},t) = \sqrt{\frac{\hbar}{l}} \sum_{\mathbf{k}} i\sqrt{\frac{\omega(\mathbf{k})}{2}} \left(-a_k e^{i(\mathbf{k}\mathbf{r}-\omega(\mathbf{k})t)} + a_k^\dagger e^{-i(\mathbf{k}\mathbf{r}-\omega(\mathbf{k})t)} \right),$$

$$(2.131)$$

and the commutation rules read as follows:

$$[\mathbf{\Phi}(\mathbf{r},t),\mathbf{\Phi}(\mathbf{r}',t)] = [\mathbf{\Pi}(\mathbf{r},t),\mathbf{\Pi}(\mathbf{r}',t)] = \mathbf{O}, \qquad (2.132)$$

$$[\mathbf{\Phi}(\mathbf{r},t),\mathbf{\Pi}(\mathbf{r}',t)] = i\hbar\delta(\mathbf{r}-\mathbf{r}')\mathbf{I}. \qquad (2.133)$$

2.7 Time evolution operator and the postulate of quantum dynamics

The universe evolves in time and so does the physical system. Any quantum state in the quantum mechanical system always changes in time, just like the motion of the particles in classical mechanics changes its configuration according to laws of motion. All the physical states as well as the dynamical observables we have dealt with in the first three postulates of quantum mechanics were expressed in such a way that time is totally absent from the equations. In fact, the physical quantities we had met or dealt with are time dependent implicitly, namely that we set our observation of the system at a convenient time scale, say $t = 0$. Choosing the time scale is entirely arbitrary . We would like to express the physical state at the particular time $t = t_0$ to perform the observation or to make the measurement. Then the state will be denoted by $|\psi; t_0\rangle$. All the contents in our previous discussion do not change and all the theories and conclusions we have derived so far remain the same if we replace $|\psi\rangle$ by $|\psi; t_0\rangle$.

2.7.1 Time evolution operator and Schrödinger equation

What is more interesting and challenging for us is how the state $|\psi; t_0\rangle$ will change from the moment t_0 to any future time t. Let us denote a time evolution operator by $\mathbf{U}(t, t_0)$. It is a linear operator whose action upon a physical state $|\psi; t_0\rangle$ will evolve the state into a new state $|\psi; t\rangle$ governed by the law of quantum dynamics. Mathematically it is expressed by

$$|\psi; t\rangle = \mathbf{U}(t, t_0)|\psi; t_0\rangle. \qquad (2.134)$$

Conservation of the norm square at any time, or conservation of the probability when they are expressed in the q-representation, requires that the time evolution operator be unitary, namely,

$$\langle\psi;t|\psi;t\rangle = \langle\psi;t_0|U^\dagger(t,t_0)U(t,t_0)|\psi;t_0\rangle = \langle\psi;t_0|\psi;t_0\rangle \quad (2.135)$$

or

$$\int \langle\psi;t|x\rangle dx \langle x|\psi,t\rangle = \int |\psi(x;t)|^2 dx = \int |\psi(x;t_0)|^2 dx.$$

Therefore

$$U^\dagger(t,t_0)U(t,t_0) = I \quad \text{or} \quad U^\dagger(t,t_0) = U^{-1}(t,t_0). \quad (2.136)$$

Let us investigate a state at time t, which evolves into state at time $t + \delta t$, an infinitesimal time interval δt after t, i.e.

$$|\psi;t+\delta t\rangle = U(t+\delta t,t)|\psi;t\rangle. \quad (2.137)$$

Since $U(t,t) = I$, the time evolution operator $U(t+\delta t,t)$, of course is very close to identity operator, can be approximated up to first order in δt by the following expression

$$U(t+\delta t,t) = I - i\Omega\delta t, \quad (2.138)$$

where the operator Ω is Hermitian because

$$U^{-1}(t+\delta t,t) = I + i\Omega^\dagger\delta t = I + i\Omega\delta t,$$

or $\Omega^\dagger = \Omega$.

It is the operator Ω that dictates the change of the state, and this forms the last postulate of quantum mechanics, namely the law of quantum dynamics, in which Ω is taken to be $H(P,Q;t)/\hbar$.

4th Postulate of Quantum Mechanics:

The evolution of the quantum state is governed by the following dynamical law,

$$|\psi; t + \delta t\rangle = \mathbf{U}(t, t + \delta t)|\psi; t\rangle = \left(\mathbf{I} - \frac{i}{\hbar}\mathbf{H}(\mathbf{P}, \mathbf{Q}; t)\delta t\right)|\psi; t\rangle$$
$$(2.139)$$

or to rewrite it in the familiar expression of the Schrödinger equation,

$$i\hbar \lim_{\delta t \to 0} \frac{1}{\delta t}(|\psi; t + \delta t\rangle - |\psi; t\rangle) = i\hbar\frac{\partial}{\partial t}|\psi; t\rangle = \mathbf{H}(\mathbf{P}, \mathbf{Q}; t)|\psi; t\rangle.$$
$$(2.140)$$

In other words, the total energy of the system is the generator that drives the quantum state into evolution. Therefore the Schrödinger equation is referred to as the law of quantum dynamics.

2.7.2 Time order product and construction of time evolution operator

The product of two time evolution operators is again a time evolution operator, given by

$$\mathbf{U}(t, t_0) = \mathbf{U}(t, t')\mathbf{U}(t', t_0).$$
$$(2.141)$$

Hence we have

$$\mathbf{U}(t + \delta t, t_0) = \mathbf{U}(t + \delta t, t)\mathbf{U}(t, t_0) = \left(\mathbf{I} - \frac{i}{\hbar}\mathbf{H}(\mathbf{P}, \mathbf{Q}; t)\right)\delta t. \quad (2.142)$$

From which the differential equation for $\mathbf{U}(t, t_0)$ is obtained as follows

$$\frac{\partial \mathbf{U}(t, t_0)}{\partial t} = \lim_{\delta t \to 0} \frac{1}{\delta t}(\mathbf{U}(t + \delta t, t_0) - \mathbf{U}(t, t_0)) = -\frac{i}{\hbar}\mathbf{H}(\mathbf{P}, \mathbf{Q}; t)\mathbf{U}(t, t_0).$$
$$(2.143)$$

The solution for $\mathbf{U}(t, t_0)$ is trivial for the conserved system in which the Hamiltonian operator does not depend upon time explicitly. $\mathbf{U}(t, t_0)$ then takes the following expression

$$\mathbf{U}(t, t_0) = e^{-\frac{i}{\hbar}\mathbf{H}(\mathbf{P},\mathbf{Q})(t-t_0)}. \tag{2.144}$$

If the Hamiltonian operator becomes time dependent, we can solve the evolution operator by iteration, i.e.

$$\mathbf{U}(t, t_0) = \mathbf{I} + \left(-\frac{i}{\hbar}\right) \int_{t_0}^{t} \mathbf{H}(\mathbf{P}, \mathbf{Q}; t')\mathbf{U}(t', t_0)dt',$$

or with the abbreviation $\mathbf{H}(\mathbf{P}, \mathbf{Q}; t) = \mathbf{H}(t)$, then

$$\mathbf{U}(t, t_0) = \mathbf{I} + \left(-\frac{i}{\hbar}\right) \int_{t_0}^{t} dt'\mathbf{H}(t') + \left(-\frac{i}{\hbar}\right)^2 \int_{t_0}^{t} dt'\mathbf{H}(t') \int_{t_0}^{t'} dt''\mathbf{H}(t'') + \dots.$$

Let us define the time order product as follows

$$\mathcal{T}(\mathbf{A}(t_1)\mathbf{B}(t_2)) = \begin{cases} \mathbf{A}(t_1)\mathbf{B}(t_2) & \text{if } t_1 > t_2, \\ \mathbf{B}(t_2)\mathbf{A}(t_1) & \text{if } t_2 > t_1, \end{cases}$$

then

$$\int_{t_0}^{t} dt'\mathbf{H}(t') \int_{t_0}^{t'} dt''\mathbf{H}(t'') = \frac{1}{2!} \iint_{t_0}^{t} dt'dt''\mathcal{T}(\mathbf{H}(t')\mathbf{H}(t'')),$$

and $\mathbf{U}(t, t_0)$ can be cast into more compact expression as

$$\mathbf{U}(t, t_0) = \mathbf{I} + \left(-\frac{i}{\hbar}\right) \int_{t_0}^{t} dt' \mathbf{H}(t') + \frac{1}{2!} \left(-\frac{i}{\hbar}\right)^2 \iint_{t_0}^{t} dt' dt'' \mathcal{T}(\mathbf{H}(t')\mathbf{H}(t''))$$

$$+ \frac{1}{3!} \left(-\frac{i}{\hbar}\right)^3 \iiint_{t_0}^{t} dt' dt'' dt''' \mathcal{T}(\mathbf{H}(t')\mathbf{H}(t'')\mathbf{H}(t''')) + \cdots$$

$$= \mathcal{T} \left\{ e^{-\frac{i}{\hbar} \int_{t_0}^{t} dt \mathbf{H}(\mathbf{P}, \mathbf{Q}; t)} \right\}. \tag{2.145}$$

2.8 Schrödinger picture vs. Heisenberg picture

Measurement of the physical quantities differs from time to time in the dynamical system. The expectation values of the dynamical observables in a quantum system correspond to the data of the measurements, whose values depend upon the time at which the measurements are performed. Therefore, when we take the expectation value of a dynamical observable $\mathbf{A}$ with respect to the state $|\psi; t\rangle$, i.e. the value $\langle \mathbf{A} \rangle = \langle \psi; t | \mathbf{A} | \psi; t \rangle$ that we obtain, in fact, is time dependent. From now on we shall denote the implicitly time dependent expectation value of the dynamical operator $\mathbf{A}$ with respect to the state $|\psi; t\rangle$ by

$$\langle \mathbf{A} \rangle_t = \langle \psi; t | \mathbf{A} | \psi; t \rangle. \tag{2.146}$$

Taking the time derivative of the above equation, we have the following relation

$$\frac{d}{dt} \langle \mathbf{A} \rangle_t = \frac{d}{dt} \langle \psi; t | \mathbf{A} | \psi; t \rangle = \frac{d}{dt} \langle \psi; t_0 | \mathbf{U}^\dagger(t, t_0) \mathbf{A} \mathbf{U}(t, t_0) | \psi; t_0 \rangle$$

$$= \frac{i}{\hbar} \langle \psi; t_0 | \mathbf{U}^\dagger (\mathbf{H}\mathbf{A} - \mathbf{A}\mathbf{H}) \mathbf{U} | \psi; t_0 \rangle = \frac{i}{\hbar} \langle \psi; t | [\mathbf{H}, \mathbf{A}] | \psi; t \rangle. \tag{2.147}$$

If we take $\mathbf{A}$ to be the position operator $\mathbf{Q}$, the equation becomes

$$\frac{d}{dt}\langle \mathbf{Q} \rangle_t = \frac{i}{\hbar} \langle \psi; t | [\mathbf{H}, \mathbf{Q}] | \psi; t \rangle. \qquad (2.148)$$

The commutator of $\mathbf{H}$ and $\mathbf{Q}$ can be evaluated according to the fundamental commutation relations among coordinate and momentum, and we obtain

$$[\mathbf{H}, \mathbf{Q}] = \frac{\hbar}{i} \frac{\partial \mathbf{H}}{\partial \mathbf{P}},$$

hence

$$\frac{d}{dt}\langle \mathbf{Q} \rangle_t = \left\langle \frac{\partial \mathbf{H}}{\partial \mathbf{P}} \right\rangle_t. \qquad (2.149)$$

Similarly in the case for the momentum operator $\mathbf{P}$,

$$\frac{d}{dt}\langle \mathbf{P} \rangle_t = -\left\langle \frac{\partial \mathbf{H}}{\partial \mathbf{Q}} \right\rangle_t. \qquad (2.150)$$

It is in this Schrödinger picture that treating the quantum state as time dependent while keeping the dynamical observables as fixed quantum operators, Hamilton's equations of motion are regained as in the following table:

Quantum equations of motion in Schrödinger picture	classical equations of motion
$\dfrac{d}{dt}\langle \mathbf{Q} \rangle_t = \left\langle \dfrac{\partial \mathbf{H}}{\partial \mathbf{P}} \right\rangle_t,$	$\dot{q}_c(t) = \dfrac{\partial H_c}{\partial p_c(t)},$
$\dfrac{d}{dt}\langle \mathbf{P} \rangle_t = -\left\langle \dfrac{\partial \mathbf{H}}{\partial \mathbf{Q}} \right\rangle_t,$	$\dot{p}_c(t) = -\dfrac{\partial H_c}{\partial q_c(t)},$

where $H_c = H(p_c(t), q_c(t))$.

There also exists another picture to look at the quantum equations of motion, namely the Heisenberg picture. Instead of regarding the state vector as time dependent, one takes the new dynamical observables by the unitary transformation of the old ones, namely taking the new dynamical operator as

$$\tilde{\mathbf{A}}(t) = \mathbf{U}^\dagger(t,0)\mathbf{A}\mathbf{U}(t,0) = \mathbf{U}^\dagger(t)\mathbf{A}\mathbf{U}(t), \qquad (2.151)$$

here for simplicity and clarity, we take $t_0 = 0$ and put $\mathbf{U}(t,t_0) = \mathbf{U}(t)$.

Since the commutation relations are invariant under the unitary transformation, the fundamental quantization rules remain the same as before, provided that all new operators are of at equal time, i.e.

$$[\tilde{\mathbf{P}}(t), \tilde{\mathbf{Q}}(t)] = \frac{\hbar}{i}\mathbf{I}, \qquad (2.152)$$

which allows us to establish the following commutation relations

$$[\tilde{\mathbf{Q}}(t), F(\tilde{\mathbf{P}}(t), \tilde{\mathbf{Q}}(t))] = -\frac{\hbar}{i}\frac{\partial F}{\partial \tilde{\mathbf{P}}(t)}, \qquad (2.153)$$

$$[\tilde{\mathbf{P}}(t), G(\tilde{\mathbf{P}}(t), \tilde{\mathbf{Q}}(t))] = \frac{\hbar}{i}\frac{\partial G}{\partial \tilde{\mathbf{Q}}(t)}. \qquad (2.154)$$

Then the quantum equations of motion take the following expression

$$\frac{\partial \tilde{\mathbf{Q}}(t)}{\partial t} = \frac{i}{\hbar}\mathbf{U}^\dagger(t)[\mathbf{H}, \mathbf{Q}]\mathbf{U}(t) = \frac{\partial \mathbf{H}}{\partial \tilde{\mathbf{P}}(t)}, \qquad (2.155)$$

$$\frac{\partial \tilde{\mathbf{P}}(t)}{\partial t} = \frac{i}{\hbar}\mathbf{U}^\dagger(t)[\mathbf{H}, \mathbf{P}]\mathbf{U}(t) = -\frac{\partial \mathbf{H}}{\partial \tilde{\mathbf{Q}}(t)}, \qquad (2.156)$$

and we have the following table:

Quantum equations of motion in Heisenberg picture	classical equations of motion
$\dfrac{\partial \tilde{\mathbf{Q}}(t)}{\partial t} = \dfrac{\partial \mathbf{H}}{\partial \tilde{\mathbf{P}}(t)},$	$\dot{q}_c(t) = \dfrac{\partial H_c}{\partial p_c(t)},$
$\dfrac{\partial \tilde{\mathbf{P}}(t)}{\partial t} = -\dfrac{\partial \mathbf{H}}{\partial \tilde{\mathbf{Q}}(t)},$	$\dot{p}_c(t) = -\dfrac{\partial H_c}{\partial q_c(t)}.$

2.9 Propagator in quantum mechanics

Let us consider a one-dimensional system in which the Hamiltonian does not depend on time explicitly. The time evolution operator therefore takes the simple form,

$$\mathbf{U}(t, t_0) = e^{-\frac{i}{\hbar}\mathbf{H}(\mathbf{P},\mathbf{Q})(t-t_0)} = U(t - t_0).$$

The wave function in the Schrödinger picture can be expressed as

$$\psi(x; t) = \langle x|\psi; t\rangle = \langle x|\mathbf{U}(t - t')|\psi; t'\rangle$$

$$= \int \langle x|e^{-\frac{i}{\hbar}\mathbf{H}(t-t')}|x'\rangle dx' \langle x'|\psi; t'\rangle = \int K(x, t; x', t')\psi(x', t')dx',$$

and the equality of the last equation defines the **propagator**

$$K(x, t; x', t') = \langle x|e^{-\frac{i}{\hbar}\mathbf{H}(t-t')}|x'\rangle. \tag{2.157}$$

It is in fact the propagator that is taken as the matrix element of the time evolution operator $\mathbf{U}(t, t')$ between the states $|x\rangle$ and $|x'\rangle$. The propagator brings the state function at spacetime point (x', t') to the state function at another spacetime point (x, t). It is obvious that when $t = t'$, the propagator reduces to the delta function, i.e.

$$K(x, t; x', t) = \langle x|e^{-\frac{i}{\hbar}\mathbf{H}(t-t)}|x'\rangle = \langle x|x'\rangle = \delta(x - x'). \tag{2.158}$$

We shall explore the property of the propagator by putting it in a different form. Let us insert an identity projection operator formed by energy eigenstates in front of state $|x'\rangle$, then we express the propagator in the following form

$$K(x,t;x',t') = \sum_\alpha \langle x|e^{-\frac{i}{\hbar}\mathbf{H}(t-t')}|\alpha\rangle\langle\alpha|x'\rangle = \sum_\alpha \langle x|e^{-\frac{i}{\hbar}E_\alpha(t-t')}|\alpha\rangle\langle\alpha|x'\rangle.$$

$$(2.159)$$

The time derivative of the propagator can be performed as

$$i\hbar\frac{\partial}{\partial t}K(x,t;x',t') = \sum_\alpha E_\alpha\langle x|\alpha\rangle e^{-\frac{i}{\hbar}E_\alpha(t-t')}\langle\alpha|x'\rangle$$

$$= \sum_\alpha \mathbf{H}\left(\frac{\hbar}{i}\frac{\partial}{\partial x},x\right)\langle x|\alpha\rangle e^{-\frac{i}{\hbar}E_\alpha(t-t')}\langle\alpha|x'\rangle$$

$$= \mathbf{H}\left(\frac{\hbar}{i}\frac{\partial}{\partial x},x\right)K(x,t;x',t').$$

$$(2.160)$$

If we introduce the step function

$$\eta(t-t') = \begin{cases} 1, & \text{for } t > t', \\ 0, & \text{for } t < t', \end{cases}$$

and make use of the property

$$\frac{d}{dt}\eta(t-t') = \delta(t-t'),$$

then the time derivative for the product of $\eta(t-t')$ and $K(x,t;z',t')$ can be calculated as

$$i\hbar\frac{\partial}{\partial t}(\eta(t-t')K(x,t;x',t)) = \eta(t-t')\mathbf{H}K + i\hbar\delta(t-t')K$$

$$= \eta(t-t')\mathbf{H}K + i\hbar\delta(t-t')\delta(x-x'),$$

$$(2.161)$$

where the last term is obtained by the property that $\delta(x - a)f(x) = \delta(x - a)f(a)$. Hence

$$\left(i\hbar\frac{\partial}{\partial t} - \mathbf{H}\right)\eta(t - t')K(x, t; x', t') = i\hbar\delta(t - t')\delta(x - x'). \quad (2.162)$$

Therefore the propagator is also referred to as Green's function or kernel.

A simple demonstration is in order to evaluate the propagator for a free particle system in which we take $|\alpha\rangle = |p\rangle$, and summation over α is replaced by integration over the momentum p, then

$$K(x, t; x', t') = \int \langle x|p\rangle e^{-\frac{i}{\hbar}\frac{p^2}{2m}(t-t')}\langle p|x'\rangle dp$$

$$= \frac{1}{2\pi\hbar} \int e^{\frac{i}{\hbar}p(x-x')}e^{-\frac{i}{\hbar}\frac{p^2}{2m}(t-t')}dp$$

$$= \sqrt{\frac{m}{2\pi i\hbar(t - t')}} \exp\left[\frac{i}{2\hbar}\frac{m(x - x')^2}{(t - t')}\right].$$

This formula reminds us the familiar form of the delta function

$$\delta(x - a) = \lim_{\epsilon\to 0} \sqrt{\frac{1}{\pi\epsilon}} \exp\left[-\frac{(x - a)^2}{\epsilon}\right], \quad (2.163)$$

and $\lim_{t\to t'} K(x, t; x', t') = \delta(x - x')$ is reproduced.

2.10 Newtonian mechanics regained in the classical limit

Modern physics is treated usually as a new branch of physics. New theories are developed to analyze the new discoveries and phenomena in physical systems, particularly in the microscopic systems of atoms or sub-atoms. Quantum theory rose after the end of the 19th century and is commonly believed to be a correct theory for investigating the phenomena that were discovered in microscopic quantum systems in the modern era. It is a new theory that replaced Newtonian theory which was successful and widely accepted in the studies of classical dynamics.

Yet if we carefully examine the propagator we have treated in the last section, and take the very same theory in the limit that $\hbar$ tends to zero, we are able to recover the whole of the theories developed in classical dynamics.

In order to treat the spatial coordinates and the temporal ones on equal footing, we introduce a new notation as follows

$$|x, t\rangle = e^{\frac{i}{\hbar}\mathbf{H}t}|x\rangle. \tag{2.164}$$

The time dependent state function and the Schrödinger equation we had before can be re-expressed respectively as

$$\psi(x; t) = \langle x, t|\psi\rangle = \langle x|e^{-\frac{i}{\hbar}\mathbf{H}t}|\psi\rangle, \tag{2.165}$$

$$i\hbar\frac{\partial}{\partial t}\psi(x; t) = i\hbar\frac{\partial}{\partial t}\langle x|e^{-\frac{i}{\hbar}\mathbf{H}t}|\psi\rangle = \mathbf{H}\langle x, t|\psi\rangle = \mathbf{H}\psi(x; t). \tag{2.166}$$

Furthermore, the propagator from the initial spacetime point (x_0, t_0) to final spacetime point (x, t) takes the following inner product,

$$\langle x, t|x_0, t_0\rangle = K(x, t; x_0, t_0) = \langle x|e^{-\frac{i}{\hbar}\mathbf{H}(t-t_0)}|x_0\rangle, \tag{2.167}$$

which in fact is the transition probability for a quantum object at spacetime (x_0, t_0) to another spacetime (x, t). Since the closure relation holds at any instant, we have,

$$\int |x', t'\rangle dx'\langle x', t'| = \mathbf{I},$$

that can be inserted in between the state $\langle x, t|$ and the state $|x_0, t_0\rangle$ at the instant t'. If we consider the case that the transition takes place in N steps from the initial spacetime (x_0, t_0) to the final spacetime (x, t), and let

$$t - t_0 = N\epsilon, \quad t_n = t_0 + n\epsilon, \quad \text{where } n = 1, 2, \ldots, N - 1,$$

with $x_N = x$ and $t_N = t$ as shown in Figure 2.2, then the propagator can be expressed as in the following form

$$\langle x, t | x_0, t_0 \rangle = \int \prod_i dx_i \langle x, t | x_{N-1}, t_{N-1} \rangle \ldots \langle x_n, t_n | x_{n-1}, t_{n-1} \rangle \ldots$$

$$\times \langle x_1, t_1 | x_0, t_0 \rangle. \tag{2.168}$$

A typical term $\langle x_n, t_n | x_{n-1}, t_{n-1} \rangle$ in the above formula shall be explored in detail as follows

$$\langle x_n, t_n | x_{n-1}, t_{n-1} \rangle = \langle x_n | e^{-\frac{i}{\hbar} \mathbf{H} \epsilon} | x_{n-1} \rangle = \langle x_n | e^{-\frac{i}{\hbar} (\frac{1}{2m}) \mathbf{P}^2 + V(\mathbf{X})) \epsilon} | x_{n-1} \rangle$$

$$= \int \langle x_n | e^{-\frac{i}{2m\hbar} \mathbf{P}^2 \epsilon} | \xi \rangle d\xi \langle \xi | e^{-\frac{i}{\hbar} V(\mathbf{X}) \epsilon} | x_{n-1} \rangle. \tag{2.169}$$

We have made the approximation in the last equality by neglecting the term higher than order ϵ^2. The term $\langle x_n | \exp \left(-\frac{i}{2m\hbar} \mathbf{P}^2 \epsilon \right) | \xi \rangle$ can be

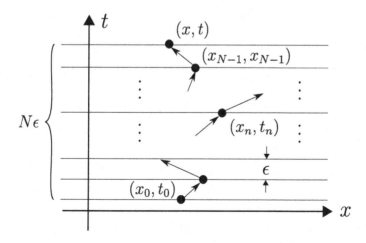

Figure 2.2: Spacetime diagram.

calculated explicitly by Fourier transform as in the case of $K(x, t; x', t')$ for a free particle system and we obtain,

$$\langle x_n | e^{-\frac{i}{2m\hbar}\mathbf{P}^2\epsilon} | \xi \rangle = \int \langle x_n | e^{-\frac{i}{2m\hbar}\mathbf{P}^2\epsilon} | p \rangle dp \langle p | \xi \rangle$$

$$= \frac{1}{2\pi\hbar} \int e^{-\frac{i}{2m\hbar}p^2\epsilon} e^{\frac{i}{\hbar}p(x_n-\xi)} dp = \sqrt{\frac{m}{2\pi i\hbar\epsilon}} \exp\left[i\frac{m(x_n-\xi)^2}{2\hbar\epsilon} \right].$$

The typical term $\langle x_n; t_n | x_{n-1}, t_{n-1} \rangle$ can finally be cast into the expression,

$$\langle x_n; t_n | x_{n-1}, t_{n-1} \rangle = \sqrt{\frac{m}{2\pi i\hbar\epsilon}} \exp\left\{ \frac{i}{\hbar}\epsilon\left[\frac{m}{2}\left(\frac{x_n - x_{n-1}}{\epsilon} \right)^2 - V(x_{n-1}) \right] \right\}$$

$$= \sqrt{\frac{m}{2\pi i\hbar\epsilon}} \exp\left(\frac{i}{\hbar}\epsilon L_n \right), \qquad (2.170)$$

where the discrete form of Lagrangian L_n takes the following form

$$L_n = \frac{m}{2}\left(\frac{x_n - x_{n-1}}{\epsilon} \right)^2 - V(x_{n-1}),$$

and the propagator of Eq. (2.167) is eventually evaluated as

$$\langle x, t | x_0, t_0 \rangle = \left(\frac{m}{2\pi i\hbar\epsilon} \right)^{N/2} \int \prod dx_n \exp\left(\frac{i}{\hbar}\epsilon L_n \right).$$

Instead of performing the multiple integration for all steps, we let N tend to infinity, and the summation over the discrete Lagrangian in the exponent becomes integration over time. This new way of performing the integration has a special name called **path integration**, after Feynman. Therefore we express

$$\langle x, t | x_0, t_0 \rangle = \lim_{\epsilon \to 0} \left(\frac{m}{2\pi i \hbar \epsilon} \right)^{N/2} \int \prod dx_n \exp \left(\frac{i}{\hbar} \epsilon L_n \right)$$

$$= \int \mathcal{D}x(\tau) \exp \left(\frac{i}{\hbar} \int_{t_0}^{t} L(x(\tau), \dot{x}(\tau)) d\tau \right), \qquad (2.171)$$

where ϵ and x_n (or x_{n-1}) are replaced by $d\tau$ and $x(\tau)$ respectively, and

$$L(x(\tau), \dot{x}(\tau)) = \lim_{\epsilon \to 0} L_n = \lim_{\epsilon \to 0} \left\{ \frac{m}{2} \left(\frac{x_n - x_{n-1}}{\epsilon} \right) - V(x_{n-1}) \right\}.$$

In the classical limit when $\hbar \to 0$, the path integration over the exponential function would oscillate so rapidly and the propagator, the transition probability, would become negligibly small. If we regard the initial spacetime as the location of a particle at the beginning time t_0, the particle ends up at the final spacetime, the final position x at time t, along the path determined by laws of classical dynamics, then we shall get the maximum transition probability. This can only be achieved if the phase in the exponential appearing in the integrant of the path integration is most stable, namely the variation of the phase be zero. Mathematically it is expressed by

$$\delta S = \delta \int_{t_0}^{t} L(x(\tau), \dot{x}(\tau)) d\tau, \qquad (2.172)$$

which is exactly the principle of least action in classical mechanics that leads to Lagrange's equations of motion, namely

$$\frac{d}{d\tau} \left(\frac{\partial L}{\partial \dot{x}(\tau)} \right) - \frac{\partial L}{\partial x(\tau)} = 0, \qquad (2.173)$$

and classical mechanics is regained.

2.11 Exercises

Ex 2.11.1

Show that delta function $\delta(x)$ can be expressed as

(a) $\delta(x) = \dfrac{1}{\pi} \lim\limits_{N \to \infty} \dfrac{\sin Nx}{x}$ or

(b) $\delta(x) = \dfrac{1}{2} \dfrac{d^2}{dx^2} |x|.$

Ex 2.11.2

Show that

$$\langle x|\mathbf{P}^2|x'\rangle = \left(\frac{\hbar}{i}\right)^2 \delta''(x - x'),$$

and hence that

$$\langle x|F(\mathbf{P})|x'\rangle = F\left(\frac{\hbar}{i}\frac{d}{dx}\right)\delta(x - x').$$

Ex 2.11.3

Consider a physical system of one dimension that is translated by a distance ξ to the right, the wave function then becomes

$$\langle x|\mathbf{U}(\mathbf{P};\xi)|\psi\rangle.$$

Evaluate the translated wave function explicitly and interpret the result. (**Hint:** take adjoint conjugate of $\mathbf{U}^\dagger(\mathbf{P};\xi)|x\rangle$ as $\langle x|\mathbf{U}(\mathbf{P};\xi).$)

Ex **2.11.4**

Show that the p-representation of the Schrödinger equation can also be expressed as

$$\frac{p^2}{2m}\psi(p) + V\left(-\frac{\hbar}{i}\frac{d}{dp}\right)\psi(p) = E\psi(p).$$

Ex **2.11.5**

Solve the Schrödinger equation for the potential energy

$$V(x) = -g\delta(x-a).$$

with g positive definite, in q-representation and p-representation.

Ex **2.11.6**

Find the one dimensional quantum state of minimum uncertainty in the p-representation.

Ex **2.11.7**

Find the quantum uncertainty $\Delta\boldsymbol{\xi}\Delta\boldsymbol{p}_\xi$ in the n particle state.

Ex **2.11.8**

Show that $\langle 0|e^{ikx}|0\rangle = \exp(-\frac{k^2}{2}\langle 0|\mathbf{X}^2|0\rangle)$, where $|0\rangle$ is the ground state of the one dimensional harmonic oscillator, and $\mathbf{X}$ is the position operator.

Ex **2.11.9**

Prove that

$$\frac{1}{2m}\sum p_r p_r = \frac{\hbar}{2}\sum_s P_s P_s^\dagger,$$

$$\frac{\beta}{2}\sum q_r q_r = \frac{\beta\hbar}{2m}\sum_s Q_s Q_s^\dagger,$$

if we define

$$p_r = \sqrt{\frac{\hbar}{Nm}}\sum_s e^{\frac{2\pi i}{N}rs}P_s,$$

$$q_r = \sqrt{\frac{m\hbar}{N}}\sum_s e^{\frac{2\pi i}{N}rs}Q_s.$$

(**Hint:** by means of Proposition 9.)

Ex 2.11.10

Prove that

$$\frac{\alpha}{2}\sum_r (q_r - q_{r+1})^2 = \frac{\alpha\hbar}{2m}\sum_s 4\sin^2\left(\frac{2\pi s}{N}\right)Q_s Q_s^\dagger.$$

(**Hint:** also by Proposition 9.)

Ex 2.11.11

By using Mehler's formula

$$\frac{1}{\sqrt{1-\xi^2}}\exp\left[-\frac{x^2+y^2-2xy\xi}{\sqrt{1-\xi^2}}\right] = e^{-(x^2+y^2)}\sum_n \frac{\xi^n}{2^n n!}H_n(x)H_n(y),$$

prove that the propagator in the case of the one dimensional harmonic oscillator can be expressed as (**Hint:** use Eq. (2.159) for $|\alpha\rangle = |n\rangle$.),

$$K(x, t; x_0, t_0) = \sqrt{\frac{m\omega}{2\pi i\hbar \sin \omega(t - t_0)}}$$

$$\times \exp\left\{\left(\frac{im\omega}{2\hbar \sin \omega(t - t_0)}\right)[(x^2 + x_0^2)\cos\omega(t - t_0) - 2xx_0]\right\}.$$

Ex 2.11.12

For a dynamical observable $\mathbf{A}(t)$ which is time dependent explicitly, the time derivative of the expectation value takes the form as

$$i\hbar\frac{d}{dt}\langle\mathbf{A}(t)\rangle_t = \langle[\mathbf{A}, \mathbf{H}]\rangle_t + i\hbar\left\langle\frac{\partial\mathbf{A}}{\partial t}\right\rangle_t.$$

Use the above formula repeatedly and show that

$$\langle\Delta x\rangle_t^2 = \langle\Delta x\rangle_0^2 + \frac{2}{m}\left\{\frac{1}{2}\langle xp + px\rangle_0 - \langle x\rangle_0\langle p\rangle_0\right\}t + \frac{\langle p\rangle_0^2}{m^2}t^2,$$

for the case of a free particle in one-dimension.

Ex 2.11.13

Prove that the one dimensional Schrödinger equation in free space

$$i\hbar\frac{\partial}{\partial t}\psi(x, t) = -\frac{\hbar^2}{2m}\frac{\partial^2}{\partial x^2}\psi(x, t)$$

is invariant under Galilean transformation:

$$x' = x - v_0 t,$$
$$t' - t.$$

(**Hint:** let $\psi'(x', t') = f(x, t; v_0)\psi(x, t)$.)

Ex 2.11.14

Derive explicitly the propagator in the p-representation, i.e.

$$\langle x|e^{-\frac{i}{\hbar}\mathbf{H}(t-t')}|x'\rangle = \int \mathcal{D}x\mathcal{D}p \exp\left(\frac{i}{\hbar}\int_{t'}^{t}\bar{L}(\tau)d\tau\right),$$

where

$$\mathcal{D}x\mathcal{D}p = \lim_{\epsilon\to 0}\prod_{n}\left(\frac{dx_n dp_n}{2\pi\hbar}\right), \qquad \int_{t'}^{t}\bar{L}(\tau)d\tau = \lim_{\epsilon\to 0}\epsilon\bar{L}_n,$$

with

$$\bar{L}_n = p_n\dot{x}_n - H(x_n, p_n),$$

$$\dot{x}_n = \frac{1}{\epsilon}(x_n - x_{n-1}).$$

Chapter 3

Symmetry, Transformation and Continuous Groups

3.1 Symmetry and transformation

Nature shows symmetries in physics, not only in the geometrical sense; it also contains a wide range of deeper implications among the invariance and the conservation laws in physics. Let us take space as an example. If we assume the homogeneity in spatial property, a translation along any direction in this space will not change the description of the system. The physical quantities involved will be invariant under the spatial transformation, or briefly, it is said to be translational invariant, and hence the total momentum of the system is conserved. So as to the isotropy property of the space, the total angular momentum is conserved due to the rotational symmetry.

Transformations in physics often form a group which is characterized by a set of continuous parameters, called the group parameters. Henceforth we call it the **continuous group**. We shall start with some simple examples in the following section.

3.1.1 Groups and group parameters

Let us consider the translation in n-dimensional space. A vector x in $\mathcal{R}^n$-space is transformed into a vector x' by a translation specified with n continuous parameters denoted by the contravariant components $\xi = (\xi^1, \xi^2, \ldots, \xi^n)$. We shall write the transformation as

$$x' = x + \xi \quad \text{or} \quad x'^i = x^i + \xi^i \quad \text{for} \quad i = 1, 2, \ldots, n. \qquad (3.1)$$

Further translation from the vector x' to the vector x'' by another displacement η which is characterized by another set of n continuous parameters $\eta = (\eta^1, \eta^2, \ldots, \eta^n)$ is written as

$$x'' = x' + \eta \quad \text{or} \quad x''^i = x'^i + \eta^i = x^i + (\xi^i + \eta^i). \qquad (3.2)$$

It is obvious that all the translations form a group, called the n-dimensional **translational group**. If we denote the group elements simply by ξ and η for the previous two consecutive translations, and put a dot $\cdot$ in between ξ and η, then the group operation for the consecutive translation is expressed as

$$\gamma = \eta \cdot \xi,$$

and the composition rules for the group parameters are just the addition, i.e.

$$\gamma = (\gamma^1, \gamma^2, \ldots, \gamma^n), \quad \gamma^i = \xi^i + \eta^i. \qquad (3.3)$$

Hence the transformation of the space translation satisfies the group postulates.

(a) γ is again a translation with the group element specified by the set of n continuous parameters, i.e. $x'' = x + \gamma$, $\gamma^i = \xi^i + \eta^i$.

(b) There exists the identity element $e = (0, 0, \ldots, 0)$ such that $e \cdot \xi = \xi \cdot e = \xi$.

(c) The inverse element of ξ, denoted by ξ^{-1} exists, such that

$$\xi^{-1} \cdot \xi = \xi \cdot \xi^{-1} = e,$$

that implies

$$\xi^{-1} = (-\xi^1, -\xi^2, \ldots, -\xi^n). \qquad (3.4)$$

Instead of performing the displacement upon the vectors in $\mathcal{R}^n$-space, if we consider the transformation of the function $F(x)$ by making the displacement in the coordinates system, then the expression for the group elements will be drastically changed. When the function $F(x) = F(x^1, x^2, \ldots, x^n)$ is transformed by a displacement into $F(x') = F(x + \xi) = F(x^1 + \xi^1, x^2 + \xi^2, \ldots, x^n + \xi^n)$, we can expand the function in terms of a Taylor series and put it in a more compact expression by summing over the repeated indices as

$$F(x') = F(x + \xi)$$

$$= F(x) + \xi^i \frac{\partial}{\partial x^i} F(X) + \frac{1}{2!} \xi^i \xi^j \frac{\partial}{\partial x^i} \frac{\partial}{\partial x^j} F(x) + \ldots = e^{\xi^i \frac{\partial}{\partial x^i}} F(x).$$

(3.5)

One notices that the group elements for translation take a completely different form. The group element of the displacement is written as $\exp\left(\xi^i \partial/\partial x^i\right)$ instead of ξ. Therefore the group operation is the usual multiplication, namely the successive displacements of ξ and η is written as the product of two group elements $\exp\left(\eta^i \partial/\partial x^i\right)$ and $\exp\left(\xi^i \partial/\partial x^i\right)$, and reaching to another element as

$$\exp\left(\eta^i \frac{\partial}{\partial x^i}\right) \cdot \exp\left(\xi^i \frac{\partial}{\partial x^i}\right) = \exp\left[(\xi^i + \eta^i)\frac{\partial}{\partial x^i}\right] = \exp\left(\gamma^i \frac{\partial}{\partial x^i}\right),$$

and the addition of the group parameters $\gamma^i = \xi^i + \eta^i$ is regained in the exponent but in totally different context. Namely the fact that the partial derivatives with respect to two different coordinates commute with each other ensure us to have $\gamma^i = \xi^i + \eta^i$.

The parameters in the translational group form an n-dimensional Euclidean space itself. This $\mathcal{R}^n$-space is called the group parameter space, or **group manifold** in short.

Let us consider the linear transformation on n-dimensional vector space as another example. Vector x when expressed by a column matrix, is changed into vector x' by an $n \times n$ matrix A, i.e.

$$x' = Ax,$$

or if we express it in terms of the matrix element as

$$x'^i = A^i_j x^j = (\delta^i_j + a^i_j)x^j, \quad \text{where} \quad A^i_j = \delta^i_j + a^i_j.$$

The linear transformations form a group provided that all the transformation matrices are of nonvanishing determinants. Consider two successive transformations and $x' = Ax$, $x'' = Bx'$, then we conclude:

(a) $x'' = Bx' = BAx = Cx$ and $C = BA$ with property det $C \neq 0$ because det $A \neq 0$ and det $B \neq 0$.

(b) The identity matrix exists by choosing $a_j^i = 0$.

(c) Since det $A \neq 0$, of course A^{-1} exists.

The group element A of this linear transformation consists of n^2 parameters, namely a_j^i for $i, j = 1, 2, \ldots, n$. Therefore its group parameter space or the group manifold is a $\mathcal{R}^{n^2}$-space, a Euclidean space of n^2-dimension, and this group is called the $GL(n, \mathcal{R})$ group.

The composition rules for the group parameters in $GL(n, \mathcal{R})$ are more complicated than in the case of spatial translation. The group parameters of the group element $C = BA$ obtained from the successive transformations A and B take the following composition rules

$$c_j^i = (\delta_l^i + b_l^i)(\delta_j^l + a_j^l), \tag{3.6}$$

or

$$c_j^i = \delta_j^i + b_j^i + a_j^i + b_l^i a_j^l. \tag{3.7}$$

Different groups dictate different sets of the composition rules for the group parameters. We shall consider the general case of r-parameter group and denote the element simply by the set of r parameters. Let us denote the group elements by

$$a, b \quad \text{for} \quad a = (a^1, a^2, \ldots, a^r), \ b = (b^1, b^2, \ldots, b^r).$$

For the product of a and b, i.e. $c = ba = f(a^i, b^j) = f(a, b)$, we can express the group parameters of the resultant element $c^l = f^l(a^i, b^j)$ with the composition functions $f^l(a^i, b^j)$ being infinitively differentiable with respect to a^i and b^j.

Denote e as the identity element and identify its position at the origin in the parameter space, i.e. $e = (0, 0, \ldots, 0)$, then we have

$$f(a, e) = f(e, a) = a. \tag{3.8}$$

Also if we denote the inverse of element a by a^{-1}, then the following relations hold

$$f(a, a^{-1}) = f(a^{-1}, a) = e \quad \text{or} \quad f^l(a, a^{-1}) = f^l(a^{-1}, a) = 0. \tag{3.9}$$

Continuous groups were systematically investigated by Lie who particularly concentrated on the behavior of function $f(a, b)$ when both elements a and b are at the vicinity of the identity transformation, namely putting elements a and b close to identity element e, and looking into the functional behavior of $f^l(a^i, b^j)$ for infinitesimal of a^i and b^j. We shall summarize the important results in the next section.

3.2 Lie groups and Lie algebras

Let us start with the composition rules of the group parameters

$$c = f(a, b) \quad \text{or} \quad c^l = f^l(a^i, b^j).$$

Since the elements $a = (a^1, a^2, \ldots, a^r)$ and $b = (b^1, b^2, \ldots, b^r)$ are close to the identity element $e = (0, 0, \ldots, 0)$, we perform Taylor's expansion of the functions $f^l(a^i, b^j)$ at the origin of the group parameter space in power series of a^i and b^j up to second order, namely,

$$c^l = f^l(a^i, b^j) = f^l(e, e) + \frac{\partial f^l(e, e)}{\partial a^i} a^i + \frac{\partial f^l(e, e)}{\partial b^i} b^i + \frac{\partial^2 f^l(e, e)}{\partial a^i \partial b^j} a^i b^j + \mathcal{O}(3).$$

This equation can be further simplified by the following relations:

$$f^l(e, e) = 0, \quad f^l(a, e) = a^l, \quad f^l(e, b) = b^l,$$

and hence

$$c^l = f^l(a^i, b^j) = a^l + b^l + f^l_{ij} a^i b^j + \mathcal{O}(3),$$

with

$$f^l_{ij} = \frac{\partial^2 f^l(e, e)}{\partial a^i \partial b^j} = \text{constant coefficient.} \tag{3.10}$$

The above formula enables us to calculate the group parameters of the inverse element $a^{-1} = ((a^{-1})^1, (a^{-1})^2, \ldots, (a^{-1})^n)$. When $a \sim e$, we evaluate $f(a, a^{-1})$ as

$$0 = c^l = f^l(a, a^{-1}) = a^l + (a^{-1})^l + f_{ij}^l a^i (a^{-1})^j + \mathcal{O}(3),$$

therefore, up to O(3) we have,

$$(a^{-1})^l = -a^l + f_{ij}^l a^i a^j + \mathcal{O}(3).$$

Let us consider the transformation of a vector in n-dimensional space, as shown in the Figure 3.1, such that the transformation forms a group $G(a)$, i.e.

$$x \xmapsto{G(a)} x' = h(x, a),$$

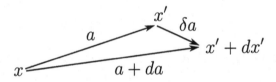

Figure 3.1: Transformation of a vector in n-dimensional space.

where $G(a)$ is the r-parameter group, and a is the group element.

Let us perform an infinitesimal transformation from x' to x'' by δa, an element close to identity element e, right after the first transformation by a, as follows

$$x'' = x' + dx' = h(x', \delta a).$$

The changes in the vector can be expressed as

$$dx' = \frac{\partial h(x', e)}{\partial(\delta a)} \delta a = u(x') \delta a,$$

where $u(x')$ stands for a $n \times r$ matrix.

The combined transformation can also be obtained through a single transformation by the group element $a + da$ with the following relations:

$$x'' = h(x, a + da), \quad a + da = f(a, \delta a).$$

Therefore da can also be expressed as the first order approximation in δa as follows

$$da = \frac{\partial f(a, e)}{\partial b}\delta a, \quad \text{or} \quad da^l = \frac{\partial f^l(a, e)}{\partial b^m}\delta a^m = f^l_m(a)\delta a^m, \qquad (3.11)$$

where $\partial f(a, e)/\partial b$ is a $r \times r$ square matrix evaluated at $b = e$ or if we express δa in terms of da by inverting the matrix $\partial f(a, e)/\partial b$, namely

$$\delta a = \left(\frac{\partial f(a, e)}{\partial b}\right)^{-1} da = \psi(a)da, \qquad (3.12)$$

where $\psi(a)$ again is a $r \times r$ square matrix, the inverse matrix of $\partial f(a, e)/\partial b$. Therefore

$$dx' = u(x')\psi(a)da,$$

or when it is written in terms of the components as follows

$$dx'^m = u^m_i \psi^i_j da^j.$$

Let us consider the transformation for an arbitrary function $F(x) = F(x^1, x^2, \ldots, x^n)$ in the vector space by a group element at the vicinity of identity e. The change in $F(x)$ is calculated as

$$dF(x) = \frac{\partial F(x)}{\partial x}dx = \frac{\partial F(x)}{\partial x}u(x)\delta a = \delta a^i \left(u^m_i \frac{\partial}{\partial x^m}\right)F(x) = \delta a^i X_i F(x).$$

We define the differential operator

$$X_i = u^m_i \frac{\partial}{\partial x^m}, \qquad (3.13)$$

which is called the generator of the transformation, or the **infinitesimal generator** of the group, or simply **group generator**.

Let us go back to the space translation discussed in previous chapter. A vector x in $\mathcal{R}^3$-space is displaced by a group element $\xi = (\xi^1, \xi^2, \xi^3)$, namely

$$x \xmapsto{T(\xi)} x' = x + \xi,$$

or expressed in components

$$x'^m = x^m + \xi^m.$$

The generators of the 3-dimensional translational group can be calculated as follows

$$X_i = u_i^m \frac{\partial}{\partial x^m} = \frac{\partial x'^m}{\partial \xi^i} \frac{\partial}{\partial x^m} = \delta_i^m \frac{\partial}{\partial x^m} = \frac{\partial}{\partial x^i},$$

which is related to the i-th component of the momentum operators in q-representation in quantum mechanics, i.e.

$$\frac{\partial}{\partial x^i} = \frac{i}{\hbar} P_i, \tag{3.14}$$

which allows us to express $U(P, -\xi) = e^{\xi^i \frac{\partial}{\partial x^i}} = e^{\frac{i}{\hbar} \vec{\xi} \cdot \vec{P}}$ as the group element for a coordinate transformation to the right by $x^i \to x^i + \xi^i$, which corresponds to the transformation of the physical state to the left by ξ^i.

We take the rotation about the azimuthal axis as another example. It is a one-parameter group defined by

$$x'^1 = x^1 \cos\theta - x^2 \sin\theta, \tag{3.15}$$

$$x'^2 = x^1 \sin\theta + x^2 \cos\theta, \tag{3.16}$$

that allows us to calculate

$$u_1^1 = \left.\frac{\partial x'^1}{\partial \theta}\right|_{\theta=0} = -x^2, \tag{3.17}$$

$$u_1^2 = \left.\frac{\partial x'^2}{\partial \theta}\right|_{\theta=0} = x^1, \tag{3.18}$$

and we obtain the generator of the rotational group about z-axis as follows

$$X_3 = u_1^i \frac{\partial}{\partial x^i} = -x^2 \frac{\partial}{\partial x^1} + x^1 \frac{\partial}{\partial x^2} = \frac{i}{\hbar} L_3, \tag{3.19}$$

which is a familiar expression related to the q-representation of the 3rd component of the angular momentum operator.

The angle θ of the rotation in fact is just the group parameter. The parameter takes the value $0 \leqslant \theta < 2\pi$. We call it the O(2) group, an orthogonal transformation in $\mathcal{R}^2$-space. A continuous group with a finite number of parameters that takes a bounded domain in the group parameter space is called **compact Lie group**. Otherwise we call them non-compact ones. The translational group in the previous example is a non-compact 3-parameter group, while the O(2) group is a compact 1-parameter group.

Lie algebra is defined as the commutator of the group generators. Let us consider the commutator of X_i and X_j, i.e.

$$[X_i, X_j] = [u_i^l \frac{\partial}{\partial x^l}, u_j^m \frac{\partial}{\partial x^m}] = u_i^l \frac{\partial u_j^m}{\partial x^l} \frac{\partial}{\partial x^m} - u_j^m \frac{\partial u_i^l}{\partial x^m} \frac{\partial}{\partial x^l}$$

$$= \left(u_i^l \frac{\partial u_j^m}{\partial x^l} - u_j^l \frac{\partial u_i^m}{\partial x^l} \right) \frac{\partial}{\partial x^m}. \tag{3.20}$$

Yet the last parenthesis can be simplified as

$$u_i^l \frac{\partial u_j^m}{\partial x^l} - u_j^l \frac{\partial u_i^m}{\partial x^l} = c_{ij}^k u_k^m, \tag{3.21}$$

where $c_{ij}^k = -c_{ji}^k$ is a constant coefficient and anti-symmetric with respect to i and j.

The proof, though a little tedious, goes as follows:

Since $dx^m = u_k^m \psi_n^k da^n$, we have $\dfrac{\partial x^m}{\partial a^n} = u_k^m \psi_n^k$. With the property of infinite differentiability of the transformation functions $h(x, a)$ with respect to the group parameters, namely that

$$\frac{\partial^2 x^m}{\partial a^l \partial a^n} = \frac{\partial^2 x^m}{\partial a^n \partial a^l}, \tag{3.22}$$

we reach at

$$u_k^m \left(\frac{\partial \psi_l^k}{\partial a^n} - \frac{\partial \psi_n^k}{\partial a^l} \right) + \frac{\partial u_k^m}{\partial a^n} \psi_l^k - \frac{\partial u_k^m}{\partial a^l} \psi_n^k = 0. \tag{3.23}$$

Putting $\dfrac{\partial u_k^m}{\partial a^n} = \dfrac{\partial u_k^m}{\partial x^l} u_\alpha^l \psi_n^\alpha$, the above equation takes the expression as

$$u_k^m \left(\frac{\partial \psi_l^k}{\partial a^n} - \frac{\partial \psi_n^k}{\partial a^l} \right) + \frac{\partial u_k^m}{\partial x^\alpha} u_\beta^\alpha \psi_n^\beta \psi_l^k - \frac{\partial u_k^m}{\partial x^\alpha} u_\beta^\alpha \psi_l^\beta \psi_n^k = 0,$$

or cast into the following expression

$$u_k^m \left(\frac{\partial \psi_l^k}{\partial a^n} - \frac{\partial \psi_n^k}{\partial a^l} \right) = \left(\frac{\partial u_\beta^m}{\partial x^k} u_\alpha^k - \frac{\partial u_\alpha^m}{\partial x^k} u_\beta^k \right) \psi_l^\alpha \psi_n^\beta. \tag{3.24}$$

By inverting ψ into f, the above equation can be written as

$$u_k^m \left(\frac{\partial \psi_l^k}{\partial a^n} - \frac{\partial \psi_n^k}{\partial a^l} \right) f_i^l f_j^n = \left(\frac{\partial u_j^m}{\partial x^k} u_i^k - \frac{\partial u_i^m}{\partial x^k} u_j^k \right). \qquad (3.25)$$

The term on the right hand side of the last equation depends upon the coordinate only, while the left hand side is product of a coordinate dependent function u_k^m and a group parameter dependent factor $\left(\frac{\partial \psi_l^k}{\partial a^n} - \frac{\partial \psi_n^k}{\partial a^l} \right) f_i^l f_j^n$. The condition can only be satisfied if,

$$\left(\frac{\partial \psi_l^k}{\partial a^n} - \frac{\partial \psi_n^k}{\partial a^l} \right) f_i^l f_j^n = c_{ij}^k = -c_{ji}^k = \text{ constant coefficient.}$$

Therefore we have,

$$\frac{\partial u_j^m}{\partial x^k} u_i^k - \frac{\partial u_i^m}{\partial x^k} u_j^k = c_{ij}^k u_k^m, \qquad (3.26)$$

and hence the commutator of X_i and X_j can then be expressed as

$$[X_i, X_j] = \left(u_i^k \frac{\partial u_j^m}{\partial x^k} - u_j^k \frac{\partial u_i^m}{\partial x^k} \right) \frac{\partial}{\partial x^m} = c_{ij}^k u_k^m \frac{\partial}{\partial x^m} = c_{ij}^k X_k, \quad (3.27)$$

and they are called **Lie algebras**. c_{ij}^k are called the **structure constants**.

A few terminologies concerning Lie groups are introduced as follows:

(a) A Lie group is said to be Abelian if all the structure constants are zero, i.e.

$$c_{ij}^k = 0, \quad i, j, k \leqslant r,$$

which imply that the generators of the group commute with each other.

(b) A subset of the elements of a group G is called a subgroup of G if those elements satisfy the group postulates. The generators of the subgroup, $X_1, X_2, \ldots, X_i, \ldots, X_p$ for $p < r$ are closed, namely they form a sub-Lie algebra by themselves, and

$$[X_i, X_j] = c_{ij}^k X_k, \quad \text{with } c_{ij}^k = 0 \text{ if } i, j \leqslant p, \ k > p.$$

(c) It is called an **invariant subgroup** H of G, if the generators in H commute with the generators outside H, namely that

$$[X_i, X_j] = c_{ij}^k X_k, \quad \text{with } c_{ij}^k = 0 \text{ if } i \leqslant p, \ k > p.$$

(d) A group is called a simple group if it contains no invariant subgroup. It is called a semi-simple if it contains no invariant, Abelian subgroup.

The generators of the group also satisfy the **Jacobi identity**, i.e.

$$[[X_i, X_j], X_k] + [[X_j, X_k], X_i] + [[X_k, X_i], X_j] = 0, \tag{3.28}$$

which leads to the following relation on the structure constants,

$$c_{ij}^l c_{lk}^m + c_{jk}^l c_{li}^m + c_{ki}^l c_{lj}^m = 0. \tag{3.29}$$

3.3 More on semisimple group

We shall confine ourselves to semisimple Lie group from now on for further investigation. A symmetric g tensor is introduced for the discussion on Cartan's criteria of the semisimple group, and also the inverse of the g tensor, i.e. g^{-1}, another symmetric one will be constructed for the discussion on the Casimir operator.

3.3.1 Cartan's criteria of the semisimple group

With the structure constants we construct a symmetric tensor g defined as

$$g_{ij} = c_{im}^l c_{jl}^m = g_{ji}, \tag{3.30}$$

which is called the metric tensor or g tensor or sometimes by the name of Killing form. We formulate the criteria given by Cartan in characterizing the semisimple group in the following proposition.

Proposition 1.

A group G is semisimple if and only if $det|g| \neq 0$.

The condition $det|g| \neq 0$ for a semisimple group is necessary and sufficient. Let us assume a Lie group contains an Abelian invariant subgroup. We denote the index of the generators in the Abelian invariant subalgebra by putting a bar on the top, i.e. $\bar{i}$, $\bar{j}$ and $\bar{k}$. It can be easily demonstrated that $det|g|$ of the group vanishes, i.e.

$$det|g| = 0, \tag{3.31}$$

if the group contains an Abelian invariant subgroup. Let us calculate the element $g_{i\bar{j}}$ of the g tensor as follows

$$g_{i\bar{j}} = c_{im}^{l} c_{\bar{j}l}^{m} = c_{i\bar{m}}^{l} c_{\bar{j}l}^{\bar{m}} = c_{i\bar{m}}^{\bar{l}} c_{\bar{j}\bar{l}}^{\bar{m}} = 0, \tag{3.32}$$

which implies that the elements of the whole column of the g tensor equal to zero. Hence the determinant of the g tensor is zero, i.e.

$$det|g| = 0,$$

and the sufficient condition of Cartan's criteria is reached.

The condition of zero in determinant of the g tensor is also necessary. Consider a system of r simultaneous homogeneous equations as follows

$$g_{ij}x^{j} = 0. \tag{3.33}$$

The condition $det|g| = 0$ also ensures the existence of the nontrivial solutions of x^{j}. Let us construct a generator X, with the coefficients taken from the nontrivial solutions x^{j}. The generators then form an invariant subgroup as demonstrated in the following proposition.

Proposition 2.

If x^j are the nontrivial solutions of the following simultaneous homogeneous equations,

$$g_{ki}x^i = 0$$

and X_i is the generators of the group G, then the generators $x^i X_i$ form an invariant subalgebra of G.

The proof is quite straightforward by considering the following commutator

$$[x^i X_i, X_j] = x^i c_{ij}^k X_k = y^k X_k, \qquad (3.34)$$

where $y^k = x^i c_{ij}^k$.

If we contract g_{ik} with y^k, we obtain the following result

$$g_{ik}y^k = g_{ik}x^l c_{lj}^k = x^l c_{lji} = c_{ji}^n g_{nl} x^l = 0, \qquad (3.35)$$

which implies that y^k is also the solution of Eq. (3.33), and hence the generators $x^i X_i$ form an invariant subalgebra. While in reaching Eq. (3.35), we have made use of the cyclic properties of the indices in $c_{lji} = c_{jil}$ which will be shown later in Proposition 3.

3.3.2 Casimir operator

The nonzero determinant of g, i.e. $\det|g| \neq 0$, enables us to construct the inverse of g, a tensor denoted by g^{-1}, with elements g^{ij} such that

$$g^{il}g_{lj} = \delta_j^i,$$

then g^{-1} will allow us to form a quadratic of the generators of the Lie group, called **Casimir operator** defined as

$$C = g^{ij}X_i X_j, \qquad (3.36)$$

which shall commute with any generator of the group as can be verified by taking the commutator of C with, say, the generator X_k in the following expression

$$[C, X_k] = g^{ij}[X_i X_j, X_k] = g^{ij}c_{ik}^l(X_l X_j + X_j X_l). \tag{3.37}$$

Let us define a 3rd rank tensor through the following relation

$$c_{ik}^l = g^{lj}c_{jik}.$$

We shall prove the anti-symmetric property of c_{ijk} in the following proposition.

Proposition 3.

The 3rd rank tensor c_{ijk} is anti-symmetric with respect to the interchange of any pair of the indices.

The proof goes as follows

$$c_{ijk} = c_{ij}^l g_{lk} = c_{ij}^l c_{lm}^n c_{kn}^m$$

$$= -c_{jm}^l c_{li}^n c_{kn}^m - c_{mi}^l c_{lj}^n c_{kn}^m = c_{jm}^l c_{il}^n c_{kn}^m + c_{mi}^l c_{lj}^n c_{nk}^m. \tag{3.38}$$

By summing over the repeated indices, the last line of the above equation becomes invariant under the cyclic permutation of i, j and k. Therefore the conclusion of Proposition 3 is reached. Eqation (3.37) can then be expressed as

$$[C, X_k] = g^{ij}g^{lm}c_{mik}(X_l X_j + X_j X_l)$$

$$= g^{ml}g^{ij}c_{mik}(X_l X_j + X_j X_l) = 0. \tag{3.39}$$

Since the last term in the last equation is symmetric with respect to l and j, so is the upper indices m and i, yet c_{mik} is anti-symmetric with respect to the lower indices m and i, the summation leads to $[C, X_k] = 0$, which concludes that the Casimir operator of the group commutes with all the group generators. We shall make use of this property to investigate the theory of angular momentum in the following chapter.

3.4 Standard form of the semisimple Lie algebras

Following the approach of Cartan and Weyl, let us consider the following equation

$$[A, X] = \rho X, \tag{3.40}$$

where A is an arbitrary linear combination of the generators X_i with the coefficient a^i, i.e.

$$A = a^i X_i, \tag{3.41}$$

and X is another linear combination of the generators X_i with the co-efficients x^i, namely

$$X = x^i X_i, \tag{3.42}$$

such that Eq. (3.40) can be met.

We shall give a special name to X of the eigenvalue equation in Eq. (3.40) and call X the **eigengenerator**, while ρ is also called the eigenvalue for this particular form of the eigenvalue equation. If we express Eq. (3.40) explicitly in terms of the structure constants c_{ij}^k, then we have the following simultaneous homogeneous equations, i.e.

$$(a^i c_{ij}^k - \rho \delta_j^k) x^j = 0, \tag{3.43}$$

with the secular equation given by

$$\det |a^i c_{ij}^k - \rho \delta_j^k| = 0. \tag{3.44}$$

For the r-parameter Lie group, there are r roots of the solution of ρ in Eq. (3.44). We shall state the conclusions from Cartan's without providing the argument in details that if a^i is chosen such that the secular equation has the maximum numbers of distinct roots, then only the eigenvalue $\rho = 0$ becomes degenerate. Let l be the multiplicity of the degenerate roots of the secular equation, then l is said to be the rank of the semisimple Lie algebra. The eigengenerator corresponding to the degenerate eigenvalue $\rho = 0$ is denoted by H_i $(i = 1, 2, \ldots, l)$, which commutes with each other, i.e.

$$[H_i, H_j] = 0, \tag{3.45}$$

with the structure constant $c_{ij}^{\mu} = 0$ where $i, j \leqslant l$ and $\mu \leqslant r$.

The eigenvalue equations with nonzero eigenvalues read as

$$[A, E_{\alpha}] = \alpha E_{\alpha}, \qquad (3.46)$$

where we have used the eigenvalue α to label the eigengenerator as E_{α}.

Since A commutes with H_i, it allows us to express A as the linear combination of H_i, namely

$$A = \lambda^i H_i. \qquad (3.47)$$

Let us investigate the commutator of A and $[H_i, E_{\alpha}]$ as follows

$$\begin{aligned}
[A, [H_i, E_{\alpha}]] &= [A, H_i E_{\alpha}] - [A, E_{\alpha} H_i] \\
&= \alpha(H_i E_{\alpha} - E_{\alpha} H_i) = \alpha[H_i, E_{\alpha}],
\end{aligned} \qquad (3.48)$$

which implies that there exist l eigengenerators corresponding to the same eigenvalue α in the eigenvalue equation of Eq. (3.40). It is the result contradictory to the assumption that α is not degenerate. Therefore we conclude that $[H_i, E_{\alpha}]$ must be proportional to E_{α}, i.e.

$$[H_i, E_{\alpha}] = \alpha_i E_{\alpha}, \qquad (3.49)$$

where the structure constant $c_{i\alpha}^{\beta} = \alpha_i \delta_{\alpha}^{\beta}$, with α and β taking all the distinct nonzero eigenvalues. One also proves readily that

$$\alpha = \lambda^i \alpha_i \qquad (3.50)$$

by making use of Eqs. (3.46), (3.47) and (3.49).

To further our investigation, let us make use of the Jacobi identity and we find that

$$\begin{aligned}
[A, [E_{\alpha}, E_{\beta}]] &= -[E_{\alpha}, [E_{\beta}, A]] - [E_{\beta}, [A, E_{\alpha}]] \\
&= (\alpha + \beta)[E_{\alpha}, E_{\beta}],
\end{aligned} \qquad (3.51)$$

which means that $[E_{\alpha}, E_{\beta}]$ is the eigengenerators of A with eigenvalue $\alpha + \beta$ if and only if $\alpha + \beta \neq 0$. Hence we can express the commutator of E_{α} and E_{β} as another eigengenerator as follows

$$[E_{\alpha}, E_{\beta}] = N_{\alpha\beta} E_{\alpha+\beta}, \qquad (3.52)$$

with the structure constant $c_{\alpha\beta}^{\gamma} = N_{\alpha\beta}\delta_{\alpha+\beta}^{\gamma}$.

For the case that $\alpha + \beta = 0$, the commutator $[E_\alpha, E_{-\alpha}]$ shall be proportional to the linear combination of H_i. Therefore we reach the following conclusions:

$$[E_\alpha, E_{-\alpha}] = c_{\alpha-\alpha}^{i} H_i, \qquad (3.53)$$

$$[E_\alpha, E_\beta] = N_{\alpha\beta} E_{\alpha+\beta}. \qquad (3.54)$$

Let us evaluate the elements of the g tensor in the standard form and divide them into three categories, namely g_{ij}, $g_{i\alpha}$ and $g_{\alpha\beta}$, where the indices in the subscripts take the following values:

$$\begin{cases} i, j \leqslant l, \\ \alpha, \beta \text{ takes the values, } r - l \text{ in total numbers, corresponding to} \\ \text{the roots of non-zero eigenvalues.} \end{cases}$$

The evaluation of g_{ij} is the simplest as follows, for

$$g_{ij} = c_{i\beta}^{\alpha} c_{j\alpha}^{\beta} = \alpha_i \alpha_j. \qquad (3.55)$$

We shall leave them as an exercise to show that

$$g_{i\alpha} = 0. \qquad (3.56)$$

Finally let us calculate the element $g_{\alpha\beta}$ as follows

$$g_{\alpha\beta} = c_{\alpha\mu}^{\nu} c_{\beta\nu}^{\mu} = c_{\alpha\gamma}^{\alpha} c_{\beta\alpha}^{\gamma} + c_{\alpha-\alpha}^{\mu} c_{\beta\mu}^{-\alpha} + \sum_{\gamma\neq-\alpha} c_{\alpha\gamma}^{\alpha+\gamma} c_{\beta(\alpha+\gamma)}^{\gamma}, \qquad (3.57)$$

where μ, ν takes the values from 1 to l as well as all those corresponding to the nonzero eigenvalues.

With the properties of Eqs. (3.49), (3.53) and Eq. (3.54), we reach the conclusion that $g_{\alpha\beta}$ is 0 unless $\beta = -\alpha$, namely

$$g_{\alpha\beta} = 0 \quad \text{if} \quad \alpha + \beta \neq 0. \qquad (3.58)$$

Suppose that α is the eigenvalue of Eq. (3.40), but $-\alpha$ is not, then the whole column in the g tensor becomes 0 because $g_{\alpha\beta} = 0$. The determinant of g will vanish in this case, and Cartan's criteria for semisimple group will be violated. Therefore the roots α and $-\alpha$ of the secular equation Eq. (3.44) must appear in pair and $g_{\alpha-\alpha}$ will be different from 0. We shall choose the nonzero $g_{\alpha-\alpha} = 1$, due to the fact that the eigengenerator E_α is linear on both sides of the eigenvalue equation of Eq. (3.40). This degree of freedom in the normalization of g_{ij} allows us to evaluate $c^i_{\alpha-\alpha}$ as follows

$$c^i_{\alpha-\alpha} = g^{ij} c_{\alpha-\alpha j} = g^{ij} c_{j\alpha-\alpha} = g^{ij} c^\beta_{j\alpha} g_{\beta-\alpha}$$

$$= g^{ij} c^\alpha_{j\alpha} \cdot 1 = g^{ij} \alpha_j = \alpha^i. \tag{3.59}$$

Hence we can rewrite Eq. (3.53) as follows

$$[E_\alpha, E_{-\alpha}] = \alpha^i H_i. \tag{3.60}$$

Here we summarize the algebra of the semisimple group in the following standard form

$$[H_i, H_j] = 0, \tag{3.45}$$

$$[H_i, E_\alpha] = \alpha_i E_\alpha, \tag{3.49}$$

$$[E_\alpha, E_{-\alpha}] = \alpha^i H_i, \tag{3.60}$$

$$[E_\alpha, E_\beta] = N_{\alpha\beta} E_{\alpha+\beta}, \text{ if } \alpha + \beta \neq 0. \tag{3.54}$$

3.5 Root vector and its properties

Let us regard the root α as an element in l-dimensional vector space with its covariant components α_i, i.e.

$$\underline{\alpha} = (\alpha_1, \alpha_2, \ldots, \alpha_l), \tag{3.61}$$

where a bar is placed under the root, written as $\underline{\alpha}$ to indicate it as a **covariant** vector. While the **contravariant** vector, denoted by $\overline{\alpha}$, with the component α^i, which has been derived previously, can be expressed as

$$\overline{\alpha} = (\alpha^1, \alpha^2, \ldots, \alpha^l). \tag{3.62}$$

If we denote the scalar product of two roots α and β as follows

$$(\alpha, \beta) = \alpha_i \beta^i = \alpha^i \beta_i, \tag{3.63}$$

then we obtain the following proposition.

Proposition 4.

If α and β are two roots, then
$$\frac{(\alpha, \beta)}{(\alpha, \alpha)} = \frac{1}{2} \ integers. \qquad (3.64)$$

We start the proof by taking the commutator of the eigengenerators $E_{-\alpha}$ and E_γ, namely $[E_{-\alpha}, E_\gamma]$. Since we enjoy the degree of freedom to rescale the eigengenerators, the structure constants can be chosen to be one, i.e.

$$[E_{-\alpha}, E_\gamma] = E_{\gamma-\alpha}. \qquad (3.65)$$

Let us assume that γ is a root but not $\gamma + \alpha$. Then we can construct a series of roots by performing the commutator successively of $E_{-\alpha}$ and a general eigenvector $E_{\gamma-j\alpha}$ with different j, i.e.

$$[E_{-\alpha}, E_{\gamma-j\alpha}] = E_{\gamma-(j+1)\alpha}. \qquad (3.66)$$

The series of roots will start from γ and terminate at $\gamma - g\alpha$, or explicitly as follows

$$\gamma, \gamma - \alpha, \gamma - 2\alpha, \ldots, \beta, \ldots, \gamma - g\alpha, \qquad (3.67)$$

which is called α-**string**, in which the root β stands for any root in the string.

The finite length of the α-string comes from the fact that it is a r-parameter group, the finite number of roots prohibit us to perform the commutator indefinitely. Let us assume the string stops at the root $\gamma - g\alpha$, then the commutator vanishes, i.e.

$$[E_{-\alpha}, E_{\gamma-g\alpha}] = E_{\gamma-(g+1)\alpha} = 0. \qquad (3.68)$$

Now let us take the commutator of eigengenerators E_α and any generator, say $E_{\gamma-(j+1)\alpha}$, then we obtain as follows

$$[E_\alpha, E_{\gamma-(j+1)\alpha}] = \mu_{j+1} E_{\gamma-j\alpha}, \qquad (3.69)$$

where the structure constant μ_{j+1} can not be normalized further for the reason that all eigengenerators corresponding to the roots in α-string have been rescaled already. Applying the Jacobi identity for the generators E_α, $E_{-\alpha}$ and $E_{\gamma-j\alpha}$, which leads to the following relation

$$\begin{aligned}
\mu_{j+1} E_{\gamma-j\alpha} &= [E_\alpha, [E_{-\alpha}, E_{\gamma-j\alpha}]] \\
&= -[E_{-\alpha}, [E_{\gamma-j\alpha}, E_\alpha]] - [E_{\gamma-j\alpha}, [E_\alpha, E_{-\alpha}]] \\
&= -[E_{\gamma-j\alpha}, \alpha^i H_i] + \mu_j [E_{-\alpha}, E_{\gamma-(j-1)\alpha}] \\
&= \alpha^i [H_i, E_{\gamma-j\alpha}] + \mu_j E_{\gamma-j\alpha},
\end{aligned}$$

and we arrive at a recursion formula for the structure constant μ_{j+1} as follows

$$\mu_{j+1} = (\gamma, \alpha) - j(\alpha, \alpha) + \mu_j, \tag{3.70}$$

where j takes the integer with $\mu_0 = 0$.
The structure constant μ_j then can be evaluated as follows

$$\mu_j = j(\gamma, \alpha) - \frac{1}{2} j(j-1)(\alpha, \alpha). \tag{3.71}$$

Since j terminates at g, therefore $\mu_{g+1} = 0$. It follows that

$$(\gamma, \alpha) = \frac{1}{2} g(\alpha, \alpha). \tag{3.72}$$

Let us now take $\beta = \gamma - j\alpha$ as any root in the α-string, and eliminate γ in Eq. (3.72), then we reach

$$(\alpha, \beta) = \frac{1}{2}(g - 2j)(\alpha, \alpha). \tag{3.73}$$

which proves Proposition 4.

3.6 Vector diagrams

We are now in the position to investigate the graphic representations of the root vectors. As we have shown in the construction of α-string that

$$(\alpha, \beta) = \frac{1}{2}m(\alpha, \alpha), \qquad (3.74)$$

where m takes the integer. Similarly, it follows that

$$(\alpha, \beta) = \frac{1}{2}n(\beta, \beta), \qquad (3.75)$$

if the β-string is taken into consideration. Therefore we obtain the following condition

$$\cos^2 \varphi = \frac{(\alpha, \beta)(\beta, \alpha)}{(\alpha, \alpha)(\beta, \beta)} = \frac{1}{4}mn, \qquad (3.76)$$

and $\cos^2 \varphi$ takes only the value 0, 1/4, 1/2, and 1.

As we have demonstrated that the roots α and $-\alpha$ come in pair, we need only to consider the positive values of $\cos \varphi$, which lead us to take the following possible angles for φ, i.e.

$$\varphi = 0, \frac{\pi}{6}, \frac{\pi}{4}, \frac{\pi}{3}, \frac{\pi}{2}.$$

The length of the root vector α and β can then be calculated according to the various combinations in m and n as follows:

$$|\alpha| = \sqrt{(\alpha, \alpha)} = \sqrt{\frac{2(\alpha, \beta)}{m}},$$

$$|\beta| = \sqrt{(\beta, \beta)} = \sqrt{\frac{2(\alpha, \beta)}{n}}.$$

With the method introduced by Van der Waerden, we summarize the results in 5 cases as follows:

(1) for $\varphi = 0$, thus we have the trivial result that two vectors α and β are identical.

(2) for $\varphi = \pi/6, m = 1$ or 3 and $n = 3$ or 1 respectively, and the ratio of the length square, i.e. $(\beta, \beta)/(\alpha, \alpha) = 1/3$ or 3.

(3) for $\varphi = \pi/4, m = 1$ or 2 and $n = 2$ or 1 respectively, and the ratio of the length square, i.e. $(\beta, \beta)/(\alpha, \alpha) = 1/2$ or 2.

(4) for $\varphi = \pi/3, m = 1$ and $n = 1$ respectively, and the ratio of the length square, i.e. $(\beta, \beta)/(\alpha, \alpha) = 1$.

(5) for $\varphi = \pi/2$, the scalar product vanishes and the ratio of the length is undetermined.

It is easy for us to demonstrate the graphic representation in two dimensional case, namely for rank two Lie groups. The root vectors can be drawn on a two dimensional space. For the above case 2 to case 4, the vector diagrams are:

A. $\varphi = \pi/6$

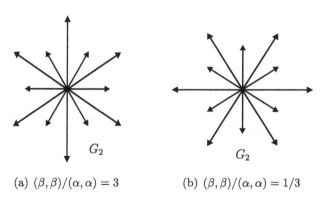

(a) $(\beta, \beta)/(\alpha, \alpha) = 3$ (b) $(\beta, \beta)/(\alpha, \alpha) = 1/3$

The group is called G_2 group after Cartan, an **exceptional group** containing 2 null root vectors and another 12 root vectors.

B. $\varphi = \pi/4$

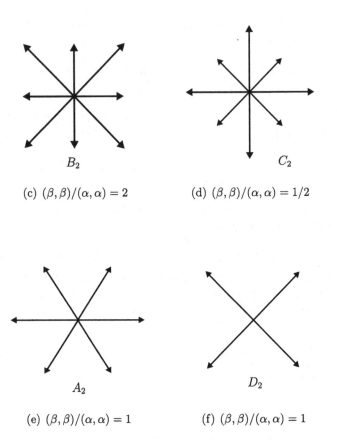

B_2

C_2

(c) $(\beta, \beta)/(\alpha, \alpha) = 2$ (d) $(\beta, \beta)/(\alpha, \alpha) = 1/2$

A_2

D_2

(e) $(\beta, \beta)/(\alpha, \alpha) = 1$ (f) $(\beta, \beta)/(\alpha, \alpha) = 1$

Figure 3.2: Vector diagram of G_2, B_2, C_2, A_2 and D_2 group.

They are called B_2 group and C_2 group, which has 2 null root vectors and 8 root vectors and are associated with SO(5) group and Sp(4) group, the symplectic group in 4-dimension, respectively.

C. $\varphi = \pi/3$

It is called A_2 group, a group associated with SU(3), which contains 6 root vectors and 2 null root vectors.

D. $\varphi = \pi/2$

We shall consider another one called D_2 group, with two pairs of mutually orthogonal root vectors. D_2 group is commonly referred to as SO(4), which is isomorphic to the direct product of two SO(3) groups.

The groups of general rank l take the names A_l, B_l, C_l and D_l corresponding to SU($l+1$), SO($2l+1$), Sp($2l$) and SO($2l$) respectively. Further investigation in those groups mentioned above as well as the exceptional Lie groups G_2, F_4, E_6, E_7 and E_8 shall be left to the readers to consult with more advanced books and journals.

3.7 PCT: discrete symmetry, discrete groups

Besides the continuous groups we have investigated in the past sections, physics involves another category of symmetry, called the discrete symmetry. The groups associated with discrete transformations are called the discrete groups. Parity, charge conjugation and time reversal transformation are among the most frequently discussed subjects in quantum mechanics, particularly in particle physics. We shall devote ourselves to study of PCT in this section.

3.7.1 Parity transformation

Let us start with parity transformation. To investigate the phenomenon of any physical system, it is imperative to use a coordinate system. Yet as far as the motion of the particle is concerned, physicists do not care if a right-handed coordinate system is used or if a left-handed one is adopted. The reason is simple. For the motion of a system of particles in the right-handed coordinate space, there always exists an identical motion in the space of its mirror image (3-dimensional one, of course). In other words, the motion of the particles is invariant under the mirror reflections. It is said that the equation of motion in classical mechanics is invariant under the parity transformation. It can be visualized as the symmetry between a physical phenomenon in space and its image phenomenon in mirror space. This is contrary to our previous understanding of the symmetry discussed in the past few sections, which are restricted solely to continuous transformations such as the spatial translation or rotation. The symmetry arising from the parity transformation on the other hand, is discrete. For this very reason, we coin a term, discrete symmetry. The immediate consequence of this discrete symmetry is that we are unable to reorient continuously a right-handed coordinate system into a left-handed one and vice versa. Therefore parity transformation falls into the category of discrete transformation.

Quantum mechanically, the parity transformation can be represented by an operator $\mathcal{P}$, which changes the position $\vec{x}$ into $-\vec{x}$, i.e.

$$\mathcal{P} : \vec{x} \xrightarrow{P.T.} \vec{x}^{\boldsymbol{p}} = -\vec{x}, \tag{3.77}$$

or in general,

$$\mathcal{P} : f(\vec{x}) \xrightarrow{P.T.} f(\vec{x}^{\boldsymbol{p}}) = f(-\vec{x}). \tag{3.78}$$

It is obvious that

$$\mathcal{P}^2 = \mathbf{I}, \tag{3.79}$$

because that two successive parity transformations leave the position vector unchanged.

The parity transformation of a dynamical observable $\mathbf{V}$ is expressed as follows

$$\mathcal{P}\mathbf{V}\mathcal{P}^{-1} = \mathbf{V}^{\boldsymbol{p}}. \tag{3.80}$$

We shall introduce two terms which are relevant to parity transformation.

An operator $\mathbf{V}$ is called polar vector operator, or simply **vector operator** if it anti-commutes with the parity operator, i.e.

$$\{\mathcal{P}, \mathbf{V}\} = 0, \tag{3.81}$$

or more explicitly that

$$\mathcal{P}\mathbf{V}\mathcal{P}^{-1} = \mathbf{V}^{\boldsymbol{p}} = -\mathbf{V}. \tag{3.82}$$

The position operator and the momentum operator are of this kind, i.e.

$$\mathcal{P}\mathbf{X}\mathcal{P}^{-1} = -\mathbf{X}, \tag{3.83a}$$

$$\mathcal{P}\mathbf{P}\mathcal{P}^{-1} = -\mathbf{P}. \tag{3.83b}$$

On the other hand, an operator $\mathbf{A}$ which commutes with the parity, is called the **axial vector operator**, namely

$$[\mathcal{P}, \mathbf{A}] = 0, \tag{3.84}$$

or

$$\mathcal{P}\mathbf{A}\mathcal{P}^{-1} = \mathbf{A}^p = \mathbf{A}. \tag{3.85}$$

One can easily prove that the angular momentum $\mathbf{L}$ is an axial vector operator.

It is of interest to investigate a quantum system with symmetry under parity transformation. We shall summarize a few remarkable results in the following proposition.

Proposition 5.

If a dynamical operator $\mathbf{S}$ is invariance under the parity transformation, then there exists a pair of vectors which are the simultaneous eigenvectors of $\mathbf{S}$ and $\mathcal{P}$ with the eigenvalues s and ± 1 respectively.

Let us consider the following eigenvalue equation

$$\mathbf{S}|s\rangle = s|s\rangle.$$

By applying the parity operator on both sides of the last equation, we can easily show that

$$\mathcal{P}\mathbf{S}|s\rangle = \mathcal{P}\mathbf{S}\mathcal{P}^{-1}\mathcal{P}|s\rangle = \mathbf{S}^p|s\rangle^p = \mathbf{S}|s\rangle^p = s|s\rangle^p, \tag{3.86}$$

where $\mathbf{S}^p = \mathbf{S}$ and $\mathcal{P}|s\rangle = |s\rangle^p$.

This implies that $|s\rangle^{\mathcal{P}}$ is also the eigenvector of $\mathbf{S}$ with the eigenvalue s, namely that the eigenvectors are of 2-fold degeneracy. Therefore it allows one to construct a pair of vectors

$$|s_\pm\rangle = \frac{1}{\sqrt{2}}(|s\rangle \pm |s\rangle^{\mathcal{P}}), \qquad (3.87)$$

with the following properties

$$\mathcal{P}|s_\pm\rangle = \pm|s_\pm\rangle, \qquad (3.88)$$

which proves Proposition 5.

If we take $\mathbf{S}$ to be the Hamiltonian operator of a central forces system, i.e.

$$\mathbf{S} = \mathbf{H} = \frac{1}{2m}\vec{P}^2 + V(|\vec{X}|). \qquad (3.89)$$

It is obvious that $\mathcal{P}\mathbf{H}\mathcal{P}^{-1} = \mathbf{H}$. Hence we have

$$i\hbar\frac{d}{dt}\langle\mathcal{P}\rangle = \langle[\mathcal{P},\mathbf{H}]\rangle = 0. \qquad (3.90)$$

Therefore the parity is conserved, or in another way of saying, symmetry is preserved under parity transformation.

3.7.2 Charge conjugation and time reversal transformation

There exists another discrete transformation called the charge conjugation. A charge conjugation is denoted by $\mathcal{C}$, which takes a particle into its antiparticle, namely $\mathcal{C}$ changes the charge of a particle q into the charge of its antiparticle $-q$. We construct a one-dimensional charge space in which all the particles take their positions on a line or an axis, the coordinate system of the charge space, according to the amount of the electric charge they bear. Particles with positive charge and particles with the negative one are respectively located on the right hand side and the left hand side of the origin of this one-dimensional frame of reference. The operation of the charge conjugation can then be regarded mathematically as mapping a point q into the mirror image point $-q$, a reflection with respect to the origin, i.e.

$$\mathcal{C} : q \overset{C.T}{\longmapsto} -q. \tag{3.91}$$

Hence the charge conjugation operator $\mathcal{C}$ is isomorphic onto the parity operator $\mathcal{P}$ of the space reflection in 1-dimensional space.

If we denote the charge operator Q associated with the physical observable of the electric charge of a particle, then the charge conjugation $\mathcal{C}$ takes Q into $-Q$ in the following similarity transformation

$$\mathcal{C} : \mathcal{C}Q\mathcal{C}^{-1} = -Q, \tag{3.92}$$

$$\text{or} \quad \{\mathcal{C}, Q\} = 0. \tag{3.93}$$

It implies that the charge conjugation operator $\mathcal{C}$ always anti-commutes with the charge operator Q.

The one-to-one correspondence between the charge conjugation and the parity transformation in one dimension allows one to apply the results derived from the parity operation to the case of the charge conjugation.

Let us now proceed to explore the third discrete transformation $\mathcal{T}$, called the time reversal transformation. We start with the following time dependent Schrödinger equation

$$i\hbar \frac{\partial}{\partial t} \psi(\vec{x}, t) = H \left(\frac{\hbar}{i} \nabla, \vec{x} \right) \psi(\vec{x}, t). \tag{3.94}$$

If we take the complex conjugate on both sides of the equation above, and replace t by $-t$, then

$$i\hbar \frac{\partial}{\partial t} \psi^*(\vec{x}, -t) = H \left(-\frac{\hbar}{i} \nabla, \vec{x} \right) \psi^*(\vec{x}, -t). \tag{3.95}$$

It occurs often that the Hamiltonian $H(-(\hbar/i)\nabla, \vec{x}) = H((\hbar/i)\nabla, \vec{x})$ if H contains only the quadratic term in $\vec{P}$. Then the Schrödinger equation becomes

$$ i\hbar\frac{\partial}{\partial t}\psi^*(\vec{x},t) = H\left(\frac{\hbar}{i}\nabla,\vec{x}\right)\psi^*(\vec{x},-t). \qquad (3.96) $$

This implies that both $\psi(\vec{x},t)$ and $\psi^*(\vec{x},-t)$ are solutions of the Schrödinger equation.

The appearance of the complex conjugate as well as the replacement of t by $-t$ in the wave function of the last equation allows one to construct the time reversal operator which bears some sort of property in antilinearity. If we decompose $\mathcal{T}$ into the product of K and T, respectively standing for an **antilinear operator** and a temporal reflection operator, then we reach the conclusions in the following proposition.

Proposition 6.

If the time reversal operator $\mathcal{T}$ takes as product of K and T, i.e.

$$ \mathcal{T} = KT, $$

$$ where \begin{cases} K = & antilinear\ operator, \\ T = & temporal\ reflection\ operator, \end{cases} $$

then both $\psi(\vec{x},t)$ and $K\psi(\vec{x},-t)$ are solutions of the time dependent Schrödinger equation as long as $KHK^{-1} = H$. Moreover, $K^2\psi = \eta\psi$ for any wave function ψ and unimodulus factor η.

The proof goes as follows. Let us consider the transformation of H and ψ,

$$ \mathcal{T}H\mathcal{T}^{-1} = KH\left(\frac{\hbar}{i}\nabla,\vec{x},-t\right)K^{-1}, $$

$$ \mathcal{T}\psi(\vec{x},t) = K\psi(\vec{x},-t). $$

Then the Schrödinger equation takes the expression as

$$i\hbar\frac{\partial}{\partial t}\boldsymbol{K}\psi(\vec{x},-t) = (\boldsymbol{K}\boldsymbol{H}\boldsymbol{K}^{-1})\boldsymbol{K}\psi(\vec{x},-t). \qquad (3.97)$$

For the stationary system with Hamiltonian containing the quadratic term in linear momentum $\vec{P}$, then $\boldsymbol{K}\boldsymbol{H}(\hbar\nabla/i,\vec{x})\boldsymbol{K}^{-1} = \boldsymbol{H}(\hbar\nabla/i,\vec{x})$. Therefore we have

$$i\hbar\frac{\partial}{\partial t}\boldsymbol{K}\psi(\vec{x},-t) = \boldsymbol{H}\left(\frac{\hbar}{i}\nabla,\vec{x}\right)\boldsymbol{K}\psi(\vec{x},-t). \qquad (3.98)$$

This proves the first part of the proposition.

If we take the inner product of any pair of vectors ψ and φ, then by the antiunitary property of the operator $\boldsymbol{K}$, we have

$$(\psi,\varphi) = (\boldsymbol{K}\varphi,\boldsymbol{K}\psi) = (\boldsymbol{K}^2\psi,\boldsymbol{K}^2\varphi) = |\eta|^2(\psi,\varphi), \qquad (3.99)$$

which implies that η is unimodulus, i.e. $|\eta| = 1$. Hence we complete the proof of Proposition 6.

An immediate consequence of the second part of Proposition 6 is the existence of Kramers degeneracy if the quantum system is invariant under time reversal transformation. We derive the result as an example.

Example

Since ψ and $\boldsymbol{K}\psi$ are solutions of the Schrödinger equation, and are also orthogonal to each other as can be proved in the following line

$$(\psi,\boldsymbol{K}\psi) = (\boldsymbol{K}^2\psi,\boldsymbol{K}\psi) = \eta(\psi,\boldsymbol{K}\psi) = -(\psi,\boldsymbol{K}\psi) = 0, \qquad (3.100)$$

by taking $\eta = -1$. Therefore the existence of Kramers degeneracy is established.

3.8 Exercises

Ex 3.8.1

Consider the following transformation,

$$x' = a^1 x + a^2.$$

(a) What are the conditions if the transformation forms a 2 parameters group?

(b) Denote the group element by $a = (a^1, a^2)$. Find the inverse element of a, namely, $a^{-1} = ((a^{-1})^1, (a^{-1})^2)$.

(c) What are the composition rules of the group parameters, i.e. try to find $c = ba$ for $c^l = c^l(a^i, b^j) = f^l(a^i, b^j)$?

(d) Can you construct the 2×2 matrix representation of the group to justify your answer in (a), (b) and (c)?

Ex 3.8.2

Consider the rotational transformation about the azimuthal axis with an angle θ, namely,

$$x'^1 = x^1 \cos \theta - x^2 \sin \theta,$$

$$x'^2 = x^1 \sin \theta + x^2 \cos \theta,$$

$$x'^3 = x^3,$$

then the function $F(x) = F(x^1, x^2, x^3)$ will be changed into

$$F(x') = F(x^1 \cos \theta - x^2 \sin \theta, x^1 \sin \theta + x^2 \cos \theta, x^3).$$

Prove explicitly that the transformed function can be obtained by

$$\exp(\theta X_3)F(x) = \exp\left[\theta\left(-x^2\frac{\partial}{\partial x^1} + x^1\frac{\partial}{\partial x^2}\right)\right]F(x^1, x^2, x^3)$$

$$= F(x'^1, x'^2, x'^3).$$

(**Hint:** use $\exp(\theta X_3) = \exp(\theta\partial/\partial\varphi)$ and write

$$F(x) = F(r\cos\varphi, r\sin\varphi, x^3).)$$

Ex 3.8.3

A conjugate subgroup is defined as $a^{-1}Ha$ if H is a subgroup, where $a \in G$. Show that the self-conjugate subgroup can alternatively be defined as the invariant subgroup.

Ex 3.8.4

Verify that the Jacobi identity leads to following relation among the structure i.e.

$$c_{ij}^l c_{lk}^m + c_{jk}^l c_{li}^m + c_{ki}^l c_{lj}^m = 0.$$

Ex 3.8.5

Show that the elements $g_{i\alpha}$ of the g tensor in the standard form of Lie algebra vanish.

Ex 3.8.6

Give the argument to verify the elements $g_{\alpha\beta}$ of the g tensor in the standard form of Lie algebra also vanish if $\alpha + \beta \neq 0$.

Ex 3.8.7

Let the parity operator be defined as (**Hint:** by using Eq. (1.39).)

$$\mathcal{P} = e^{\frac{\pi}{2}(\vec{P} \cdot \vec{X} + \vec{X} \cdot \vec{P})}.$$

Show that

$$\mathcal{P}\vec{X}\mathcal{P}^{-1} = -\vec{X},$$

$$\mathcal{P}\vec{P}\mathcal{P}^{-1} = -\vec{P}.$$

Chapter 4

Angular Momentum

4.1 O(3) group, SU(2) group and angular momentum

Rotational transformation is one of the most commonly studied subjects in physics. Rotational transformation is an orthogonal transformation, that leaves the norm of the vector invariant. Orthogonal transformations of the coordinate system of a three-dimensional vector space form a group of O(3). Yet as far as the invariance of the norm is concerned, there exists another transformation, namely a unitary transformation, the SU(2) group, which is homomorphic onto SO(3) group, and transformation also leaves the norm of a vector in $\mathcal{C}^2$-space invariant. We shall review the orthogonal transformation in three-dimensional vector space and investigate the group structure and properties of the O(3) group, as well as the SU(2) group which are utilized as the tools for the discussion of the theory of angular momentum in the following section.

4.1.1 O(3) group

Let us perform the rotational transformation in $\mathcal{R}^3$-space, i.e. the three-dimensional vector space, about a fixed direction denoted by a unit vector $\hat{n}$ as the axis of rotation with an infinitesimal angle $d\theta$. It can also be written as a vector representing the direction of the rotational axis as well as the magnitude of the rotation as expressed in the following form

$$d\theta\hat{n} = d\vec{\theta} = (d\theta^1, d\theta^2, d\theta^3), \tag{4.1}$$

where we decompose the infinitesimal rotation $d\vec{\theta}$ into its components in the last term of the above equation. For a finite rotation with an angle $\vec{\theta}$ about the fixed axis $\hat{n}$ can be also written as

$$\theta\hat{n} = \vec{\theta} = (\theta^1, \theta^2, \theta^3). \tag{4.2}$$

Henceforth, each rotation will be represented by a point at $(\theta^1, \theta^2, \theta^3)$ in the three-dimensional parameter space.

The infinitesimal rotation will change the vector $\vec{x}$ into $\vec{x}'$, expressed in the usual vectorial algebraic notation in the following form as

$$\vec{x}' - \vec{x} = d\vec{x} = d\vec{\theta} \times \vec{x}, \tag{4.3}$$

where $d\vec{\theta} = (d\theta^1, d\theta^2, d\theta^3)$ is an infinitesimal vector in the parameter space of the rotational group at the vicinity of the origin. The parameter space is also a three-dimensional Euclidean space that is isomorphic onto the previous vector space described by the coordinate system.

To express the change in the vector due to the infinitesimal rotation in terms of the vector components, one has

$$dx^i = \epsilon^i_{jk} d\theta^j x^k,$$

that allows us to calculate the matrix elements u^l_i in the rotational group O(3), i.e.

$$u^l_i = \frac{\partial x^l}{\partial \theta^i} = \epsilon^l_{ik} x^k.$$

Since $X_i = u^l_i \dfrac{\partial}{\partial x^l}$, we obtain that

$$X_i = \epsilon^l_{ik} x^k \frac{\partial}{\partial x^l}, \tag{4.4}$$

and the Lie algebra of the O(3) group is given as follows

$$[X_i, X_j] = \epsilon_{im}^l \epsilon_{jk}^n \left[x^m \frac{\partial}{\partial x^l}, x^k \frac{\partial}{\partial x^n} \right] = (\epsilon_{im}^k \epsilon_{jk}^l + \epsilon_{mj}^k \epsilon_{ik}^l) x^m \frac{\partial}{\partial x^l}$$

$$= \epsilon_{ij}^k \epsilon_{km}^l x^m \frac{\partial}{\partial x^l} = -\epsilon_{ij}^k X_k, \tag{4.5}$$

where the fourth term of the above equation is obtained by the Jacobi identity,

$$\epsilon_{im}^k \epsilon_{jk}^l + \epsilon_{mj}^k \epsilon_{ik}^l + \epsilon_{ji}^k \epsilon_{mk}^l = 0. \tag{4.6}$$

Hence we have the following Lie algebra of the rotational transformation in 3-dimensional vector space,

$$[X_i, X_j] = -\epsilon_{ij}^k X_k, \tag{4.7}$$

that reproduce the commutation relations for the quantum operators of the angular momentum, i.e.

$$[L_i, L_j] = i\hbar \epsilon_{ij}^k L_k, \tag{4.8}$$

if we put $X_i = \dfrac{i}{\hbar} L_i$.

There are various representations for the generators of the symmetry transformation as well as for the Lie algebras. The representations of the group generators and the Lie algebras we adopted so far are expressed with the purpose to investigate the properties in the symmetry transformation of the functions. We refer this representation as the canonical formulation. If we focus our attention on the transformation of the vectors directly, we attain completely different representations. Take the rotational transformation of a vector in matrix notation as follows

$$x' = \mathbf{R}x \quad \text{or} \quad \begin{pmatrix} x'^1 \\ x'^2 \\ x'^3 \end{pmatrix} = \mathbf{R} \begin{pmatrix} x^1 \\ x^2 \\ x^3 \end{pmatrix}, \tag{4.9}$$

where $\mathbf{R}$ is a 3×3 square matrix, which is called an **orthogonal matrix** because the rotational transformation leaves the norm of the vector invariant, i.e.

$$x'^T x' = x^T \mathbf{R}^T \mathbf{R} x \quad \text{or} \quad \mathbf{R}^T \mathbf{R} = \mathbf{I}. \tag{4.10}$$

Orthogonality imposes the following 6 conditions on the matrix elements of $\mathbf{R}$, i.e.

$$R^{T i}_{\ \ j} R^j_k = \delta^i_k \quad \text{or} \quad R^j_i R^j_k = \delta_{ik}. \tag{4.11}$$

The above orthogonality conditions, asides from having the following relation

$$\det \mathbf{R}^T \mathbf{R} = (\det \mathbf{R})^2 = 1,$$

will also reduce the independent parameters in the matrix $\mathbf{R}$ from 9 to 3. It is obvious that the orthogonal matrices $\mathbf{R}$ form a group and that is the very reason to name it the $O(3)$ group, the orthogonal group in 3-dimension.

To investigate the matrix representations of the group generators, let us construct the orthogonal matrix $\mathbf{R} = \mathbf{R}(\theta^1, \theta^2, \theta^3)$, where θ^1, θ^2 and θ^3 stand for the group parameters. If we take the parameters small enough and keep only the first order in θ^i, the matrix can be expressed as follows

$$\mathbf{R}(\theta^1, \theta^2, \theta^3) = \mathbf{I} + \mathbf{A}(\theta^1, \theta^2, \theta^3),$$

where the matrix $\mathbf{A}(\theta^1, \theta^2, \theta^3)$ is close to a null matrix, namely that

$$\mathbf{A}(0, 0, 0) = 0. \tag{4.12}$$

Since the orthogonality condition of the matrix $\mathbf{R}$ leads, to first order approximation, the following relation

$$\mathbf{R}^T \mathbf{R} = (\mathbf{I} + \mathbf{A}^T)(\mathbf{I} + \mathbf{A}) \simeq \mathbf{I} + \mathbf{A}^T + \mathbf{A} = \mathbf{I},$$

which implies that matrix $\mathbf{A}$ is antisymmetric The generators of the rotational group $O(3)$ are obtained by taking the derivative of matrix $\mathbf{A}(\theta^1, \theta^2, \theta^3)$ with respect to the corresponding group parameter, i.e.

$$X_i = \frac{\partial}{\partial \theta^i} \mathbf{A}(\theta^1, \theta^2, \theta^3)\Big|_{\theta^i=0}, \qquad (4.13)$$

which again is an antisymmetric matrix as well.

Let us consider the rotational transformation about the 3rd axis with an angle φ, the orthogonal matrix $\mathbf{R}(0, 0, \varphi)$ can, up to first order in φ, be approximated as

$$\mathbf{R} = \mathbf{I} + \mathbf{A}(0, 0, \varphi) = \mathbf{I} + \begin{pmatrix} 0 & -\sin\varphi & 0 \\ \sin\varphi & 0 & 0 \\ 0 & 0 & 0 \end{pmatrix} \simeq \mathbf{I} + \begin{pmatrix} 0 & -\varphi & 0 \\ \varphi & 0 & 0 \\ 0 & 0 & 0 \end{pmatrix},$$

which allows us to obtain the generator X_3 of O(3) group by taking the derivative of matrix $\mathbf{A}$ respect to φ, namely

$$X_3 = \frac{\partial \mathbf{A}}{\partial \varphi} = \begin{pmatrix} 0 & -1 & 0 \\ 1 & 0 & 0 \\ 0 & 0 & 0 \end{pmatrix}. \qquad (4.14)$$

The reason that X_3 is called the generator of the group is due to the retrievability of the rotational matrix $\mathbf{R}(0, 0, \varphi)$, even with the finite angle φ, by exponentiating the matrix φX_3 and obtaining the following result

$$\mathbf{R}(0, 0, \varphi) = e^{\varphi X_3} = \begin{pmatrix} \cos\varphi & -\sin\varphi & 0 \\ \sin\varphi & \cos\varphi & 0 \\ 0 & 0 & 1 \end{pmatrix}. \qquad (4.15)$$

Similar to the derivation of the group generator X_3, the other two generators can also be calculated and given as

$$X_2 = \begin{pmatrix} 0 & 0 & 1 \\ 0 & 0 & 0 \\ -1 & 0 & 0 \end{pmatrix}, \quad X_1 = \begin{pmatrix} 0 & 0 & 0 \\ 0 & 0 & -1 \\ 0 & 1 & 0 \end{pmatrix}, \quad (4.16)$$

by performing the infinitesimal rotations about the axis accordingly. These three matrices are referred to as the generators of O(3) group in Cartesian bases.

Let us construct the orthogonal matrix $\mathbf{R}$ in terms of the **Euler angles** φ, θ and χ. It is obtained by performing a rotation about the 3rd axis with an angle φ followed by another rotation about the 2nd axis with an angle θ and then making a third rotation about the 3rd axis with an angle χ to achieve the matrix $\mathbf{R}$ as follows

$$\mathbf{R}(\varphi, \theta, \chi) = e^{\chi X_3} e^{\theta X_2} e^{\varphi X_3}. \quad (4.17)$$

It can be easily verified that

$$\mathbf{R}(\varphi, \theta, \chi + 2m\pi) = \mathbf{R}(\varphi, \theta, \chi), \quad (4.18)$$

for m = integers. That implies only a finite domain in the parameter space is enough to exhaust all the elements of the O(3) group. Therefore the parameters take the following ranges

$$0 \leqslant \varphi \leqslant 2\pi, \quad 0 \leqslant \theta \leqslant \pi \quad \text{and} \quad 0 \leqslant \chi \leqslant \pi. \quad (4.19)$$

Furthermore the orthogonality condition is automatically built-in because of the antisymmetric property in the group generators, namely that

$$\mathbf{R}^T = e^{-\varphi X_3} e^{-\theta X_2} e^{-\xi X_3} = \mathbf{R}^{-1}. \quad (4.20)$$

It is interesting to observe that

$$\det \mathbf{R} = 1,$$

which takes only the positive root in the equation $(\det \mathbf{R})^2 - 1 = 0$. This can be understood that the matrix $\mathbf{R}(\varphi, \theta, \chi)$ is reached from the identity element $\mathbf{I} = \mathbf{R}(0, 0, 0)$ with the determinant being equal to one, by varying the group parameters continuously away from the origin of

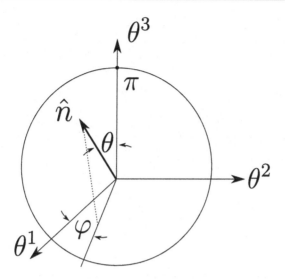

Figure 4.1: Unit vector $\hat{n}(\theta, \varphi)$.

the group parameter space. The elements of this particular orthogonal matrices form a group which is called the group of SO(3). Since the Euler angles take a finite range of the values, therefore the group parameters of SO(3) are confined within the domain of a sphere with the radius π in the parameter space. If we assign χ as the angle of rotation about a unit vector $\hat{n} = (\sin\theta\cos\varphi, \sin\theta\sin\varphi, \cos\theta)$ which serves as the axis of rotation as shown in Figure 4.1, all points within the sphere of radius π in the parameter space, representing all the elements of the rotational group, can be reached by choosing suitable $\hat{n}$ and χ, only if the two diametrically opposite points, namely points of antipode on the surface of the sphere in the group manifold, are identified.

Let us abbreviate the unit vector as $\hat{n}(\theta, \varphi)$, and denote the group generators as $\vec{X} = (X_1, X_2, X_3)$, then the group elements can be expressed as follows

$$\mathbf{R}(\varphi, \theta, \chi) = e^{\chi\hat{n}(\theta,\varphi)\cdot\vec{X}}. \tag{4.21}$$

The elements of the O(3) group consist of $\mathbf{R}(\varphi, \theta, \chi)$ and $-\mathbf{R}(\varphi, \theta, \chi)$ and $-\mathbf{R}(\varphi, \theta, \chi)$, the mirror images of the elements $\mathbf{R}(\varphi, \theta, \chi)$, namely taking spatial reflection followed by a rotation. In the other word,

O(3) group contains the elements $\mathbf{R}(\varphi, \theta, \chi)$ as well as the elements $-\mathbf{R}(\varphi, \theta, \chi)$.

4.1.2 U(2) group and SU(2) group

Let us now consider the complex vector space of 2 dimensions, i.e. C^2- space, in which a vector ξ is represented by a column matrix with two components of complex numbers as follows

$$\xi = \begin{pmatrix} \xi^1 \\ \xi^2 \end{pmatrix}. \tag{4.22}$$

A linear transformation on the vector by a matrix $\mathbf{M}$ is expressed as

$$\xi' = \mathbf{M}\xi, \quad \text{or} \quad \begin{pmatrix} \xi'^1 \\ \xi'^2 \end{pmatrix} = \begin{pmatrix} \alpha & \beta \\ \gamma & \delta \end{pmatrix} \begin{pmatrix} \xi^1 \\ \xi^2 \end{pmatrix}, \tag{4.23}$$

where the matrix elements of $\mathbf{M}$ are all of complex numbers.

If we restrict the transformation to be unitary, which leaves the norm of the vector invariant, then

$$\xi'^\dagger \xi' = \xi^\dagger \mathbf{M}^\dagger \mathbf{M}\xi = \xi^\dagger \xi, \tag{4.24}$$

and we reach that

$$\mathbf{M}^\dagger \mathbf{M} = \mathbf{I} \quad \text{or} \quad \mathbf{M}^\dagger = \mathbf{M}^{-1} \quad \text{and} \quad |\det \mathbf{M}|^2 = 1. \tag{4.25}$$

The transformation stated above is called the U(2) group. If we take $\det \mathbf{M} = 1$ as an extra restriction on matrix $\mathbf{M}$, then matrix $\mathbf{M}$ is not only unitary but also unimodular, and the transformations form the group of SU(2).

From now on we shall confine ourselves in discussing SU(2), in which the condition $\mathbf{M}^\dagger = \mathbf{M}^{-1}$ can be expressed explicitly as follows

$$\begin{pmatrix} \alpha^* & \beta^* \\ \gamma^* & \delta^* \end{pmatrix} = \frac{1}{\det \mathbf{M}} \begin{pmatrix} \delta & -\beta \\ -\gamma & \alpha \end{pmatrix}. \tag{4.26}$$

Therefore we have the following relations

$$\alpha^* = \delta, \quad \beta = -\gamma^*, \tag{4.27}$$

and the matrix $\mathbf{M}$ takes the expression as

$$\mathbf{M} = \begin{pmatrix} \alpha & \beta \\ -\beta^* & \alpha^* \end{pmatrix},$$

which contains only 3 independent parameters due to the condition $\det \mathbf{M} = |\alpha|^2 + |\beta|^2 = 1$.

Let us investigate the group property of the matrix $\mathbf{M}$ when it is close to the identity element $\mathbf{I}$, and if we put $\alpha = 1 + \xi$, then the matrix $\mathbf{M}$ can be rewritten as

$$\mathbf{M} = \mathbf{I} + \begin{pmatrix} \xi & \beta \\ -\beta^* & \xi^* \end{pmatrix} = \mathbf{I} + \mathbf{D}.$$

The **unitarity** condition of matrix $\mathbf{M}$, at the vicinity of the identity element, becomes

$$\mathbf{M}^\dagger \mathbf{M} = (\mathbf{I} + \mathbf{D}^\dagger)(\mathbf{I} + \mathbf{D}) \simeq \mathbf{I} + \mathbf{D}^\dagger + \mathbf{D} = \mathbf{I},$$

which implies $\mathbf{D}^\dagger + \mathbf{D} = 0$, or more specifically that $\xi + \xi^* = 0$. Therefore we are able to rewrite the matrix $\mathbf{M}$ in the following form

$$\mathbf{M} = \mathbf{I} + \begin{pmatrix} \frac{i}{2}c & \frac{i}{2}(a - ib) \\ \frac{i}{2}(a + ib) & -\frac{i}{2}c \end{pmatrix},$$

where a, b and c, all real, are the three parameters of the SU(2) group. The generators of the group are obtained as follows

$$Y_1 = \frac{\partial \mathbf{M}}{\partial a} = \begin{pmatrix} 0 & \frac{i}{2} \\ \frac{i}{2} & 0 \end{pmatrix}, \quad Y_2 = \frac{\partial \mathbf{M}}{\partial b} = \begin{pmatrix} 0 & \frac{1}{2} \\ -\frac{1}{2} & 0 \end{pmatrix},$$

$$Y_3 = \frac{\partial \mathbf{M}}{\partial c} = \begin{pmatrix} \frac{i}{2} & 0 \\ 0 & -\frac{i}{2} \end{pmatrix}. \tag{4.28}$$

We can verify immediately that the commutators among the generators take the general expression

$$[Y_i, Y_j] = -\epsilon_{ij}^k Y_k, \tag{4.29}$$

which is exactly the same as the Lie algebras in the case of the O(3) group.

The generators of the SU(2) group are related to **Pauli matrices** as follows

$$Y_1 = \frac{i}{2}\sigma_1, \quad Y_2 = \frac{i}{2}\sigma_2, \quad Y_3 = \frac{i}{2}\sigma_3. \tag{4.30}$$

The identical Lie algebras for both groups O(3) and SU(2) do not necessary have the same domain in the group manifold. Let us construct the group elements, i.e. the matrix $\mathbf{M}$ by exponentiating the group generators as follows

$$\mathbf{M} = e^{\chi \hat{n} \cdot \vec{Y}}, \tag{4.31}$$

where $\hat{n} = \hat{n}(\theta, \varphi)$ has the same definition as that of O(3) group, yet the parameter χ takes different domain in the group manifold for the reason that both matrix $\mathbf{M}$ and matrix $-\mathbf{M}$ fulfill the unitarity condition of the transformation. Therefore the group elements of SU(2) will fill the sphere of radius 2π, namely it takes all the points in a domain of the sphere with radius 2π in the parameter space to exhaust the group elements. The parameters φ, θ and χ range as follows

$$0 \leqslant \varphi \leqslant 2\pi, \quad 0 \leqslant \theta \leqslant \pi, \quad 0 \leqslant \chi \leqslant 2\pi, \tag{4.32}$$

with again the points of antipode on the surface of the sphere are identified.

It will be left to you as an exercise to show that

$$\mathbf{M}(\chi \hat{n} \cdot \vec{Y}) = \cos \frac{\chi}{2} + i\hat{n} \cdot \vec{\sigma} \sin \frac{\chi}{2}, \tag{4.33}$$

which enables us to conclude that

$$\mathbf{M}((\chi + 2m\pi)\hat{n} \cdot \vec{Y}) = (-1)^m \mathbf{M}(\chi \hat{n} \cdot \vec{Y}). \tag{4.34}$$

While in the case of the O(3) group, in which $\mathbf{R}(\varphi, \theta, \chi + 2m\pi) = \mathbf{R}(\varphi, \theta, \chi)$, one concludes that elements $\mathbf{M}$ and $-\mathbf{M}$ in SU(2) correspond to an element $\mathbf{R}$ in O(3). It is said that the SU(2) group is homomorphic onto the O(3) group.

To obtain a deeper understanding of the homeomorphism of SU(2) onto O(3), let us construct a 2×2 matrix X defined as follows

$$\mathbf{X} = x^i \sigma_i = \begin{pmatrix} x^3 & x^1 - ix^2 \\ x^1 + ix^2 & -x^3 \end{pmatrix}. \tag{4.35}$$

A transformation in the coordinate system performed by the rotational matrix $\mathbf{R}(\varphi, \theta, \chi)$ can equivalently be achieved by the unitary transformation of the matrix $\mathbf{X}$ which is sandwiched in between the SU(2) matrix $\mathbf{M} = \exp(\chi \hat{n} \cdot \vec{Y})$ and its adjoint conjugate $\mathbf{M}^\dagger = \exp(-\chi \hat{n} \cdot \vec{Y})$ as follows

$$\mathbf{X}' = \mathbf{M}^\dagger \mathbf{X} \mathbf{M} = e^{-\chi \hat{n} \cdot \vec{Y}} \mathbf{X} e^{\chi \hat{n} \cdot \vec{Y}}. \tag{4.36}$$

The above transformation leaves the norm of the vector invariant and is justified by taking the determinant on both sides of the above equation, i.e.

$$-x'^i x'^i = \det \mathbf{X}' = (\det \mathbf{M}^\dagger)(\det \mathbf{X})(\det \mathbf{M}) = -x^i x^i |\det \mathbf{M}|^2 = -x^i x^i. \tag{4.37}$$

Let us take the rotation about the 3rd axis with angle φ as an example and evaluate the matrix $\mathbf{M} = e^{\varphi Y_3}$ explicitly as follows

$$
\mathbf{X}' = \begin{pmatrix} x'^3 & x'^1 - ix'^2 \\ x'^1 + ix'^2 & -x'^3 \end{pmatrix}
$$

$$
= \begin{pmatrix} e^{-\frac{i}{2}\varphi} & 0 \\ 0 & e^{\frac{i}{2}\varphi} \end{pmatrix} \begin{pmatrix} x^3 & x^1 - ix^2 \\ x^1 + ix^2 & -x^3 \end{pmatrix} \begin{pmatrix} e^{\frac{i}{2}\varphi} & 0 \\ 0 & e^{-\frac{i}{2}\varphi} \end{pmatrix}
$$

$$
= \begin{pmatrix} x^3 & e^{-i\varphi}(x^1 - ix^2) \\ e^{i\varphi}(x^1 + ix^2) & -x^3 \end{pmatrix}. \tag{4.38}
$$

Equating the matrix elements on both sides, we obtain

$$
x'^1 = x^1 \cos\varphi - x^2 \sin\varphi, \tag{4.39}
$$

$$
x'^2 = x^1 \sin\varphi + x^2 \cos\varphi, \tag{4.40}
$$

$$
x'^3 = x^3, \tag{4.41}
$$

and the coordinate transformation by rotation is reproduced.

4.2 O(3)/SU(2) algebras and angular momentum

Let us consider the Lie algebras of the O(3) group or the SU(2) group, in which we take the generators as follows

$$
J_i = \frac{\hbar}{i} X_i, \quad \text{or} \quad J_i = \frac{\hbar}{i} Y_i, \tag{4.42}
$$

in order to be in conformity with the notations used in quantum mechanics. The algebra then has the following usual commutation relations

$$
[J_i, J_j] = i\hbar \epsilon_{ij}^k J_k. \tag{4.43}
$$

Let us now construct Cartan's standard form of the algebra by solving the eigenvalue equation of the following type

$$[A, X] = \rho X, \tag{4.44}$$

where A is a linear combination of the generators with arbitrary choice of the coefficients a^i, and X which is also a linear combination of the generators with the unknown coefficients x^i to be determined in order to satisfy the above eigenvalue equation.

Denote A and X as follows

$$A = a^i J_i, \quad X = x^i J_i, \tag{4.45}$$

and the eigenvalue equation becomes as

$$[A, X] = i\hbar a^i x^j \epsilon_{ij}^k J_k = \rho \delta_j^k x^j J_k. \tag{4.46}$$

Therefore we reach the following set of 3 homogeneous equations as

$$(i\hbar a^i \epsilon_{ij}^k - \rho \delta_j^k) x^j = 0. \tag{4.47}$$

For the existence of the non-trivial solutions in x^j, one imposes that

$$\det(i\hbar a^i \epsilon_{ij}^k - \rho \delta_j^k) = 0, \tag{4.48}$$

or more explicitly that

$$\begin{vmatrix} -\rho & -i\hbar a^3 & i\hbar a^2 \\ i\hbar a^3 & -\rho & -i\hbar a^1 \\ -i\hbar a^2 & i\hbar a^1 & -\rho \end{vmatrix} = 0. \tag{4.49}$$

Hence, we obtain the 3 eigenvalues given as

$$\mu_0 = 0, \tag{4.50a}$$

$$\mu_+ = \hbar\sqrt{(a^1)^2 + (a^2)^2 + (a^3)^2}, \tag{4.50b}$$

$$\mu_- = -\hbar\sqrt{(a^1)^2 + (u^2)^2 + (a^3)^2}. \tag{4.50c}$$

It is the very example of 3-parameter Lie group of rank equaling to one.

If we denote J_0, J_+ and J_-, the eigengenerators corresponding respectively to the eigenvalues μ_0, μ_+ and μ_-, we find that J_0 is proportional to A. By setting

$$J_0 = \frac{a^i J_i}{\sqrt{(a^1)^2 + (a^2)^2 + (a^3)^2}} = \frac{A}{\sqrt{(a^1)^2 + (a^2)^2 + (a^3)^2}}, \qquad (4.51)$$

we obtain the following standard form for the commutator of J_0 and $J_\pm$ in Lie algebras of the O(3)/SU(2) group, $[J_0, J_\pm] = \pm\hbar J_\pm$.

Let us compute the commutator of the generators J_+ and J_-, by applying the Jacobi identity, one finds that

$$[[J_+, J_-], J_0] = 0, \qquad (4.52)$$

which implies that the commutator of J_+ and J_- is proportional to J_0, i.e.

$$[J_+, J_-] = \alpha\hbar J_0 = \hbar J_0, \qquad (4.53)$$

where we take α to be one due to the fact that the eigenvalue equation is linear in X on both sides of the equation, that allows the eigengenerators J_+ and J_- to absorb any arbitrary factor. Henceforth the Lie algebra of the O(3)/SU(2) group is summarized as follows

$$[J_0, J_\pm] = \pm\hbar J_\pm, \qquad (4.54)$$

$$[J_+, J_-] = \hbar J_0, \qquad (4.55)$$

with the hermiticity property given as

$$J_0 = J_0^\dagger, \quad J_+^\dagger = J_-, \quad J_-^\dagger = J_+. \qquad (4.56)$$

The calculation of the Casimir operator of the group, we shall leave as an exercise, and it is found that

$$C = 2g^{ij} J_i J_j. \qquad (4.57)$$

The factor 2 is introduced with the purpose to identify C with the total angular momentum operator J^2, i.e.

$$C = J^2 = J_+J_- + J_-J_+ + J_0^2. \tag{4.58}$$

Since C commutes with the generators of the group, it is then invariant under the transformation. Furthermore the eigenvalues of the Casimir operator will be used to label the dimension of the irreducible representation of the group.

Let us construct the eigenvectors of both the operators C and J_0, and denote it as $|c, \nu\rangle$, referred to commonly as the irreducible representations of the O(3)/SU(2) group, hence we have

$$C|c, \nu\rangle = c|c, \nu\rangle, \tag{4.59a}$$

$$J_0|c, \nu\rangle = \nu\hbar|c, \nu\rangle. \tag{4.59b}$$

With the commutation relations, $[J_0, J_\pm] = \pm\hbar J_\pm$, one is now ready to show the following proposition.

Proposition 1.

If $|c, \nu\rangle$ is the eigenvector of the Casimir operator C and the operator J_0 with the eigenvalues c and $\nu\hbar$ respectively, then $J_\pm|c, \nu\rangle$ are also the eigenvectors of operators C and J_0 with the eigenvalues c and $(\nu \pm 1)\hbar$ respectively.

The vectors $J_\pm|c, \nu\rangle$ can be easily proved to be also the eigenvectors of operator C because $[C, J_\pm] = 0$, which leads to the following relation

$$CJ_\pm|c, \nu\rangle = J_\pm C|c, \nu\rangle = cJ_\pm|c, \nu\rangle. \tag{4.60}$$

To show that $J_\pm|c, \nu\rangle$ are the eigenvectors of the operator J_0 with the eigenvalues $(\nu \pm 1)\hbar$, let us apply the operator J_0 upon $J_\pm|c, \nu\rangle$ and make use of the commutation relation of Eq. (4.53), then we find that

$$J_0 J_\pm|c, \nu\rangle = (J_\pm J_0 \pm \hbar J_\pm)|c, \nu\rangle = (\nu \pm 1)\hbar J_\pm|c, \nu\rangle,$$

which imply that $J_\pm|c, \nu\rangle$ are the eigenvectors of J_0 corresponding to the eigenvalues $(\nu \pm 1)\hbar$ respectively. Proposition 1 is then established.

Let us now apply Proposition 1 to generate the eigenvectors $|c, \nu\rangle$ as follows:

$$J_+|c, \nu\rangle = \alpha|c, \nu + 1\rangle, \quad J_-|c, \nu\rangle = \beta|c, \nu - 1\rangle.$$

Therefore we are able to build a sequence of vectors by applying J_+ and J_- on $|c, \nu\rangle$ consecutively, and obtain

$$|c, \nu_-\rangle, \ldots, |c, \nu-q\rangle, \ldots, |c, \nu-1\rangle, |c, \nu\rangle, |c, \nu+1\rangle, \ldots, |c, \nu+p\rangle, \ldots, |c, \nu_+\rangle.$$

The sequence must terminate somewhere on both ends, otherwise we will encounter the dilemma of having the negative norm of the eigenvector. To see this, let us take the highest positive value ν of the eigenvector, such that the norm of the vector $J_+|c, \nu_+\rangle$ vanishes, namely

$$\langle c, \nu_+|J_-J_+|c, \nu_+\rangle = \frac{1}{2}\langle c, \nu_+|(C - J_0^2 - \hbar J_0)|c, \nu_+\rangle = 0,$$

or

$$c - \hbar^2\nu_+(\nu_+ + 1) = 0, \tag{4.61}$$

where we have made use of the relation $J_-J_+ = C - J_0^2 - \hbar J_0$. Therefore one proves that the eigenvector $|c, \alpha\rangle$ with $\alpha > \nu_+$, will lead to negative norm. Similarly, $J_-|c, \nu_-\rangle = 0$, if we denote ν_- to be the lowest value of ν, and we have the relation

$$c - \hbar^2\nu_-(\nu_- - 1) = 0. \tag{4.62}$$

Solving the above two quadratic equations to obtain values of ν_+ and ν_-, we take the difference of ν_+ and ν_-, which must be an integer, because it is the total number of steps in applying J_+ or J_- to form the nonvanishing eigenvectors in the sequence. Therefore we have

$$\nu_+ - \nu_- = 2\sqrt{\frac{1}{4} + \frac{c}{\hbar^2}} - 1 = 2j, \tag{4.63}$$

where j is taken as half integers. It also allows one to express the eigenvalue of the Casimir operator as

$$c = j(j + 1)\hbar^2, \tag{4.64}$$

which helps one to calculate that

$$\nu_+ = -\nu_- = j. \tag{4.65}$$

It is the construction of the irreducible representation of the $O(3)/SU(2)$ group that enables us to quantize the dynamical observables C and J_0.

In order to express the eigenvectors of J^2 and J_0 in the usual form appearing in quantum mechanics, we shall replace vector $|c, \nu\rangle$ with the familiar form of the vector $|j, m\rangle$ from now on, and it is normalized as

$$\langle j, m|j', m'\rangle = \delta_{jj'}\delta_{mm'},$$

with $-j \leqslant m \leqslant j$, $-j' \leqslant m' \leqslant j'$.

One also has the following relations

$$J_+|j, m\rangle = \frac{\hbar}{\sqrt{2}}\sqrt{(j-m)(j+m+1)}|j, m+1\rangle, \qquad (4.66a)$$

$$J_-|j, m\rangle = \frac{\hbar}{\sqrt{2}}\sqrt{(j+m)(j-m+1)}|j, m-1\rangle. \qquad (4.66b)$$

Different values of j form different irreducible representations of the group. The 1-dimensional representation is a trivial one by taking $j = 0$. For the case that $j = 1/2$, it forms a two-dimensional representation, called the spinor representation. If we identify the vectors $|1/2, 1/2\rangle$ and $|1/2, -1/2\rangle$ with the 2-component bases column matrices respectively as follows

$$\left|\frac{1}{2}, \frac{1}{2}\right\rangle \longrightarrow \begin{pmatrix} 1 \\ 0 \end{pmatrix}, \quad \left|\frac{1}{2}, -\frac{1}{2}\right\rangle \longrightarrow \begin{pmatrix} 0 \\ 1 \end{pmatrix},$$

then the matrix representation of the generators can be constructed with the matrix element $J_{mm'}^{(1/2)} = \langle 1/2, m|J|1/2, m'\rangle$. We shall leave them as an exercise again to show that

$$J_+^{(\frac{1}{2})} = \frac{\hbar}{\sqrt{2}}\begin{pmatrix}0 & 1\\0 & 0\end{pmatrix}, \quad J_-^{(\frac{1}{2})} = \frac{\hbar}{\sqrt{2}}\begin{pmatrix}0 & 0\\1 & 0\end{pmatrix}, \quad J_0^{(\frac{1}{2})} = \frac{\hbar}{2}\begin{pmatrix}1 & 0\\0 & -1\end{pmatrix},$$

$$(4.67)$$

which allows us to cast the spin matrices as the following expression

$$S_x = \frac{1}{\sqrt{2}}(J_+^{(\frac{1}{2})} + J_-^{(\frac{1}{2})}) = \frac{\hbar}{2}\begin{pmatrix}0 & 1\\1 & 0\end{pmatrix}, \quad (4.68a)$$

$$S_y = \frac{1}{\sqrt{2}i}(J_+^{(\frac{1}{2})} - J_-^{(\frac{1}{2})}) = \frac{\hbar}{2}\begin{pmatrix}0 & -i\\i & 0\end{pmatrix}, \quad (4.68b)$$

$$S_z = J_0^{(\frac{1}{2})} = \frac{\hbar}{2}\begin{pmatrix}1 & 0\\0 & -1\end{pmatrix}. \quad (4.68c)$$

In the case of the 3-dimensional representation, of which $j = 1$, the generators can be calculated as follows:

$$J_x^{(1)} = \frac{\hbar}{\sqrt{2}}\begin{pmatrix}0 & 1 & 0\\1 & 0 & 1\\0 & 1 & 0\end{pmatrix}, \quad (4.69a)$$

$$J_y^{(1)} = \frac{\hbar}{\sqrt{2}}\begin{pmatrix}0 & -i & 0\\i & 0 & -i\\0 & i & 0\end{pmatrix}, \quad (4.69b)$$

$$J_z^{(1)} = \hbar\begin{pmatrix}1 & 0 & 0\\0 & 0 & 0\\0 & 0 & -1\end{pmatrix}, \quad (4.69c)$$

which take completely different expression from the 3 generators of the rotational group of the Cartesian form obtained in previous Subsection 4.1. In fact the generators of 3-dimensional irreducible representation are related to those of the Cartesian form through an unitary transformation which need to be elaborated more as an exercise by the readers.

4.3 Irreducible representations of O(3) group and spherical harmonics

Let us consider the irreducible representations of the O(3) group $|l, m\rangle$, in which l takes only the positive integers l and m lies in between l and $-l$, i.e. $l = 1, 2, \ldots$ and $-l \leqslant m \leqslant l$. Then the q-representation of the vector $|l, m\rangle$ in the spherical coordinate system can be expressed as $\langle \theta, \varphi | l, m \rangle$, which is exactly the spherical harmonics $Y_l^m(\theta, \varphi)$, i.e.

$$Y_l^m(\theta, \varphi) = \langle \theta, \varphi | l, m \rangle. \tag{4.70}$$

Instead of expressing the operators as $J_\pm$ and J_0, we shall denote the orbital angular momentum operators by $L_\pm$ and L_3. It is algebraically elegant to derive all the spherical harmonics of a given degree l and the order m from the spherical harmonics of the same degree l but with a lower order $m - 1$, namely building from $Y_l^m(\theta, \varphi)$ based upon the results we have obtained in last section. We shall demonstrate these relations in the following few propositions.

Proposition 2.

Let Y_l^m and Y_l^{m-1} be two spherical harmonics of the same degree l with the order m and $m-1$ respectively, then they are connected by the following relation

$$Y_l^m(\theta, \varphi)$$
$$= \frac{\sqrt{2}}{\sqrt{(l+m)(l-m-1)}} e^{i\varphi} \left(\frac{\partial}{\partial \theta} + i \cot \theta \frac{\partial}{\partial \varphi} \right) Y_l^{m-1}(\theta, \varphi). \tag{4.71}$$

The proof is straightforward. Take the q-representation of the fol lowing equation

$$\langle \theta, \varphi | l, m \rangle = \frac{\sqrt{2}}{\sqrt{(l+m)(l-m-1)}\hbar} \langle \theta, \varphi | L_+ | l, m-1 \rangle,$$

where if the q-representation of operator L_+ is expressed by

$$L_+ = e^{i\varphi} \left(\frac{\partial}{\partial \theta} + i \cot \theta \frac{\partial}{\partial \varphi} \right) \hbar, \qquad (4.72)$$

then we arrive at the conclusion of Proposition 2.

Proposition 3.

All the spherical harmonics takes the form of the product of $e^{im\varphi}$, an exponential function of φ and another function of $f(\theta)$, i.e.

$$Y_l^m(\theta, \varphi) = e^{im\varphi} f(\theta). \qquad (4.73)$$

The proof goes by taking the q-representation of the equation as follows

$$\langle \theta, \varphi | L_3 | l, m \rangle = m\hbar \langle \theta, \varphi | l, m \rangle.$$

Express the q-representation of the operator L_3 by $(\hbar/i)\partial/\partial\varphi$, then we have the following differential equation

$$\frac{\partial}{\partial \varphi} Y_l^m(\theta, \varphi) = im Y_l^m(\theta, \varphi),$$

and the solution of the above equation leads to Proposition 3, i.e.

$$Y_l^m(\theta, \varphi) = e^{im\varphi} f(\theta).$$

The property of the function $f(\theta)$ can be analyzed in the following proposition.

Proposition 4.

Function $f(\theta)$ satisfies the associated Legendre's differential equation, i.e.

$$\frac{d}{d\mu}\left\{(1-\mu^2)\frac{d}{d\mu}f(\mu)\right\} + \left\{l(l+1) - \frac{m^2}{1-\mu^2}\right\}f(\mu) = 0, \quad (4.74)$$

or more explicitly by writing $f(\theta)$ in terms of the **associated Legendre polynomial** of degree l and order m, i.e.

$$f(\theta) \propto P_l^m(\cos\theta) = P_l^m(\mu), \quad (4.75)$$

where we put $\mu = \cos\theta$, and identify $f(\theta) = f(\mu)$.

We start the proof by taking the following equation in q-representation, namely

$$\langle\theta,\varphi|L^2|l,m\rangle = \langle\theta,\varphi|l(l+1)\hbar^2|l,m\rangle.$$

Denote the q-representation of the total angular momentum operator by

$$L^2 = -\hbar^2\left\{\frac{1}{\sin\theta}\frac{\partial}{\partial\theta}\left(\sin\theta\frac{\partial}{\partial\theta}\right) + \frac{1}{\sin^2\theta}\frac{\partial^2}{\partial^2\varphi}\right\}$$

which can be evaluated by means of coordinate transformation from the Cartesian system to the spherical one. Then we reach the following equation

$$-\left\{\frac{1}{\sin\theta}\frac{\partial}{\partial\theta}\left(\sin\theta\frac{\partial}{\partial\theta}\right) + \frac{1}{\sin^2\theta}\frac{\partial^2}{\partial^2\varphi}\right\}Y_l^m(\theta,\varphi) = l(l+l)Y_l^m(\theta,\varphi).$$

Putting $\mu = \cos\theta$, the above equation can be reduced to the following associated Legendre's differential equation as

$$\frac{d}{d\mu}\left\{(1-\mu^2)\frac{d}{d\mu}f(\mu)\right\} + \left\{l(l+1) - \frac{m^2}{1-\mu^2}\right\}f(\mu) = 0,$$

and the solution of it is referred to as the associated Legendre polynomials. Therefore Proposition 4 is proved, i.e.

$$f(\mu) = A P_l^m(\mu).$$

Proposition 2 can be facilitated to construct the spherical harmonics by avoiding the painstaking methods of solving the associated Legendre's differential equation. $Y_l^m(\theta, \varphi)$ can be formulated easily by successive applications of Proposition 2 from the spherical harmonics of lowest order, namely from $Y_l^{-l}(\theta, \varphi)$. Let us first evaluate $Y_l^{-l}(\theta, \varphi)$ by taking the q-representation of the following equation

$$\langle \theta, \varphi | L_- | l, -l \rangle = 0,$$

or

$$\left(\frac{\partial}{\partial \theta} - i \cot \theta \frac{\partial}{\partial \varphi} \right) Y_l^{-l}(\theta, \varphi) = \left(\frac{\partial}{\partial \theta} - i \cot \theta \frac{\partial}{\partial \varphi} \right) e^{-il\varphi} f(\theta)$$

$$= e^{-il\varphi} \left(\frac{\partial}{\partial \theta} - l \cot \theta \right) f(\theta) = 0,$$

which leads to the solution of $f(\theta)$ as

$$f(\theta) = A \sin^l \theta.$$

The normalization constant A can be calculated by taking the inner product as follows

$$\langle l, -l | l, -l \rangle = \int |Y_l^{-l}(\theta, \varphi)|^2 d\Omega = 2\pi |A|^2 \int_0^\pi \sin^{2l+1} \theta d\theta = 1.$$

Hence A reads as

$$A = \sqrt{\frac{(2l+1)!}{4\pi} \frac{1}{2^l l!}}.$$

Therefore we are able to express $Y_l^{-l}(\theta, \varphi)$ as follows

$$Y_l^{-l}(\theta, \varphi) = \sqrt{\frac{(2l+1)!}{4\pi}} \frac{1}{2^l l!} e^{-il\varphi} \sin^l \theta.$$

The spherical harmonics $Y_l^m(\theta, \varphi)$ can then be constructed by applying the L_+ operator $l + m$ times upon $Y_l^{-l}(\theta, \varphi)$, namely

$$Y_l^m(\theta, \varphi) = \sqrt{\frac{(l-m)!}{(2l)!(l+m)!}} \left\{ e^{i\varphi} \left(\frac{\partial}{\partial \theta} + i \cot \theta \frac{\partial}{\partial \varphi} \right) \right\}^{l+m} Y_l^{-l}(\theta, \varphi)$$

$$= \frac{1}{2^l l!} \sqrt{\frac{(2l+1)(l-m)!}{4\pi(l+m)!}} \left\{ e^{i\varphi} \left(\frac{\partial}{\partial \theta} + i \cot \theta \frac{\partial}{\partial \varphi} \right) \right\}^{l+m} e^{-il\varphi} \sin^l \theta.$$

To simplify the calculation of the last term in the previous equation, we apply the identity given in Proposition 5.

Proposition 5.

Acting the operator $e^{i\varphi} \left(\frac{\partial}{\partial \theta} + i \cot \theta \frac{\partial}{\partial \varphi} \right)$ n times repeatedly upon the function $e^{ip\varphi} f(\theta)$ will result in another function of the following form, i.e.

$$\left\{ e^{i\varphi} \left(\frac{\partial}{\partial \theta} + i \cot \theta \frac{\partial}{\partial \varphi} \right) \right\}^n e^{ip\varphi} f(\theta)$$

$$= (-1)^n e^{i(p+n)\varphi} \left(\sin^{p+n} \theta \frac{d}{d \cos \theta} \sin^{-p} \theta \right) f(\theta).$$

$$(4.76)$$

It is just a pure algebraic manipulation to show the above identity. Let us take the first step to calculate the following equation

$$e^{i\varphi}\left(\frac{\partial}{\partial\theta}+i\cot\theta\frac{\partial}{\partial\varphi}\right)e^{ip\varphi}f(\theta)$$

$$=(-1)e^{i(p+1)\varphi}\left(\sin^{p+1}\theta\frac{d}{d\cos\theta}\sin^{-p}\theta\right)f(\theta).\qquad(4.77)$$

If we regard $\left(\sin^{p+1}\theta\dfrac{d}{d\cos\theta}\sin^{-p}\theta\right)f(\theta)$ as a new function of $g(\theta)$,

i.e.

$$g(\theta)=\left(\sin^{p+1}\theta\frac{d}{d\cos\theta}\sin^{-p}\theta\right)f(\theta),$$

and take the second step as follows

$$e^{i\varphi}\left(\frac{\partial}{\partial\theta}+i\cot\theta\frac{\partial}{\partial\varphi}\right)e^{i(p+1)\varphi}g(\theta)$$

$$=(-1)e^{i(p+2)\varphi}\left(\sin^{p+2}\theta\frac{d}{d\cos\theta}\sin^{-p-1}\theta\right)g(\theta)$$

$$=(-1)^2e^{i(p+2)\varphi}\left(\sin^{p+2}\theta\frac{d^2}{d^2\cos\theta}\sin^{-p}\theta\right)f(\theta).$$

Repeating n times of the same operation, we have

$$\left\{e^{i\varphi}\left(\frac{\partial}{\partial\theta}+i\cot\theta\frac{\partial}{\partial\varphi}\right)\right\}^n e^{ip\varphi}f(\theta)$$

$$=(-1)^n e^{i(p+n)\varphi}\left(\sin^{p+n}\theta\frac{d}{d\cos\theta}\sin^{-p}\theta\right)f(\theta),$$

which proves Proposition 5.

With all the preparations in the last few propositions, we are now in the position to evaluate $Y_l^m(\theta, \varphi)$ by letting $n = l + m, p = -l$ in Proposition 5, then

$$
Y_l^m(\theta, \varphi) = \sqrt{\frac{(2l+1)(l-m)!}{4\pi(l+m)!}} \frac{1}{2^l l!} \left\{ e^{i\varphi} \left(\frac{\partial}{\partial \theta} + i \cot\theta \frac{\partial}{\partial \varphi} \right) \right\}^{l+m} e^{-il\varphi} \sin^l\theta
$$

$$
= (-1)^{l+m} \sqrt{\frac{(2l+1)(l-m)!}{4\pi(l+m)!}} \frac{e^{im\varphi}}{2^l l!} \sin^m\theta \frac{d^{l+m}}{d^{l+m}\cos\theta} \sin^{2l}\theta
$$

$$
= (-1)^{l+m} \sqrt{\frac{(2l+1)(l-m)!}{4\pi(l+m)!}} \frac{e^{im\varphi}}{2^l l!} (1-\mu^2)^{\frac{m}{2}} \frac{d^{l+m}}{d^{l+m}\mu} (1-\mu^2)^l
$$

$$
= (-1)^m \sqrt{\frac{(2l+1)(l-m)!}{4\pi(l+m)!}} \frac{e^{im\varphi}}{2^l l!} P_l^m(\mu),
$$

where

$$
P_l^m(\mu) = \frac{(-1)^l}{2^l l!} (1-\mu^2)^{\frac{m}{2}} \frac{d^{l+m}}{d^{l+m}\mu} (1-\mu^2)^l = (-1)^l (1-\mu^2)^{\frac{m}{2}} \frac{d^m}{d^m\mu} P_l(\mu),
$$

$$
P_l(\mu) = \frac{1}{2^l l!} \frac{d^l}{d^l\mu} (1-\mu^2)^l,
$$

and Rodrigues' formulas for Legendre polynomials $P_l(\mu)$ as well as for the associated Legendre functions $P_l^m(\mu)$ are reproduced.

4.4 O(4) group, dynamical symmetry and the hydrogen atom

Let us investigate the orthogonal group in 4-dimensional Euclidean space, denoted by the O(4) group, in which the norm of the vector,

$$x^i x_i = \text{ invariant } \quad \text{for } i \text{ sums from 1 to 4},$$

under the orthogonal transformation.

The generators of the group in canonical form can be easily constructed by means of rotation on the (x^i, x^j)-plane similar to the case of the O(3) group as we have done before, namely, they are the 6 generators given as

$$x^i \frac{\partial}{\partial x^j} - x^j \frac{\partial}{\partial x^i}, \quad \text{for } i, j = 1, 2, 3, 4.$$

If we denote

$$M_1 = x^2 \frac{\partial}{\partial x^3} - x^3 \frac{\partial}{\partial x^2}, \quad M_2 = x^3 \frac{\partial}{\partial x^1} - x^1 \frac{\partial}{\partial x^3}, \quad M_3 = x^1 \frac{\partial}{\partial x^2} - x^2 \frac{\partial}{\partial x^1},$$

and

$$N_1 = x^1 \frac{\partial}{\partial x^4} - x^4 \frac{\partial}{\partial x^1}, \quad N_2 = x^2 \frac{\partial}{\partial x^4} - x^4 \frac{\partial}{\partial x^2}, \quad N_3 = x^3 \frac{\partial}{\partial x^4} - x^4 \frac{\partial}{\partial x^3},$$

then the Lie algebra of the O(4) group reads as

$$[M_i, M_j] = -\epsilon_{ij}^k M_k, \quad [M_i, N_j] = -\epsilon_{ij}^k N_k, \quad [N_i, N_j] = -\epsilon_{ij}^k M_k. \quad (4.78)$$

The above relations enable us to redefine the generators by the following combinations:

$$A_i = \frac{1}{2}(M_i + N_i), \quad B_i = \frac{1}{2}(M_i - N_i), \quad (4.79)$$

and the algebra for these redefined generators become

$$[A_i, A_j] = -\epsilon_{ij}^k A_k, \quad [A_i, B_j] = 0, \quad [B_i, B_j] = -\epsilon_{ij}^k B_k. \quad (4.80)$$

Among the Lie algebra of the O(4) group, we find that there exist two sets of generators A_i and B_i whose commutators respectively form a sub-algebra of the O(3) group. The O(4) algebra becomes the direct sum of two O(3) algebras and the O(4) group is then locally isomorphic onto O(3) ⊗ O(3) group.

We are now in position to discuss the hydrogen-like atoms from the dynamical symmetry properties. The Hamiltonian operator for the hydrogen-like atoms takes the following expression

$$H = \frac{1}{2m}P^2 - \frac{Ze^2}{r} = \frac{1}{2m}P^2 - \frac{k}{r}, \qquad (4.81)$$

where we have fixed the center of the Coulomb potential at the origin. The space described by this coordinate system is isotropic with respect to the rotation about any axis through the origin. It implies that the angular momentum operators are conserved because they commute with the Hamiltonian operator. Furthermore, the attractive Coulomb potential enables us to introduce another three conserved operators, called **Lenz operators**. Classically, it is the vector, called the Lenz vector or the Runge-Lenz vector, that originates from the center of the force and points to the aphelion, as shown in Figure 4.2. It is a particular feature that the orbit is closed and fixed in space for the case of the Coulomb potential. If we construct a plane perpendicular to the constant angular momentum through the origin, the closed orbit lies on the plane without

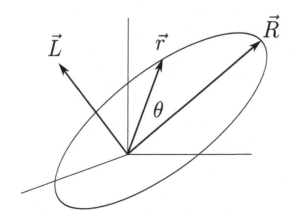

Figure 4.2: Hydrogen-like orbit.

precession of the major axis. Then the Lenz vector can be constructed
by means of the following elliptical orbit equation, i.e.

$$\frac{1}{r} = \frac{mk}{l^2}(1 - \epsilon \cos\theta), \tag{4.82}$$

where l is the angular momentum of the system and ϵ is the eccentricity
of the ellipse given as

$$\epsilon = \sqrt{1 + \frac{2El^2}{mk^2}}, \quad \text{with} \quad E < 0.$$

Let us assume the Lentz vector

$$\vec{R} = \alpha\vec{r} + \beta\vec{l} \times \vec{p}.$$

By taking $\vec{r} \cdot \vec{R}$, we reach the following relations

$$\vec{r} \cdot \vec{R} = rR\cos\theta = r\frac{l^2}{mk(1-\epsilon)}\frac{1}{\epsilon}\left\{1 - \frac{l^2}{mkr}\right\} = \alpha r^2 - \beta l^2,$$

or

$$\left(\alpha r - \frac{l^2}{mk\epsilon(1-\epsilon)}\right)r - l^2\left(\beta - \frac{l^2}{m^2k^2\epsilon(1-\epsilon)}\right) = 0,$$

that allows us to obtain

$$\alpha = \frac{1}{r}\frac{l^2}{mk\epsilon(1-\epsilon)}, \quad \beta = \frac{l^2}{m^2k^2\epsilon(1-\epsilon)},$$

which leads to the classical Lenz vector as

$$\vec{R} = \frac{l^2}{mk\epsilon(1-\epsilon)}\left\{\frac{\vec{r}}{r} + \frac{1}{mk}\vec{l} \times \vec{p}\right\}. \tag{4.83}$$

To obtain the quantum Lenz operators, we replace the classical dynamical variables by the equivalent quantum operators in the above equation. Then we have

$$R_i = \frac{l^2}{mk\epsilon(1-\epsilon)} \left\{ \frac{X_i}{\sqrt{X_1^2 + X_2^2 + X_3^2}} - \frac{1}{2mk} \epsilon_{ijk}(P_j L_k - L_j P_k) \right\}.$$
$$(4.84)$$

If we take the commutator of R_i and R_j, though it is a tedious computation, we find that

$$[R_i, R_j] = -\frac{2H}{mk^2} \frac{1}{\epsilon^2(1-\epsilon)^2} i\hbar\epsilon_{ijk}L_k.$$

By rescaling the operators in the following expression

$$\tilde{R}_i = \sqrt{\frac{mk^2}{-2H}} \epsilon(1-\epsilon)R_i,$$

the Lie algebra of the O(4) group is regained, i.e.

$$[L_i, L_j] = i\hbar\epsilon_{ijk}L_k,$$ (4.85a)

$$[L_i, \tilde{R}_j] = i\hbar\epsilon_{ijk}\tilde{R}_k,$$ (4.85b)

$$[\tilde{R}_i, \tilde{R}_j] = i\hbar\epsilon_{ijk}L_k.$$ (4.85c)

The dynamical symmetry of the Coulomb potential can also be applied to calculate the eigenenergy of the hydrogen like atoms. Let us denote

$$J_i = \frac{1}{2}(L_i + \tilde{R}_i), \quad K_i = \frac{1}{2}(L_i - \tilde{R}_i),$$ (4.86)

and these operators are the generators of O(3) ⊗ O(3) group, namely,

$$[J_i, J_j] = i\hbar\epsilon_{ijk}J_k, \tag{4.87a}$$

$$[J_i, K_j] = 0, \tag{4.87b}$$

$$[K_i, K_j] = i\hbar\epsilon_{ijk}K_k, \tag{4.87c}$$

with the property $J^2 = K^2$ because of the relation $L_i\tilde{R}_i = 0$. Let us calculate $J^2 + K^2$, which is related to the Hamiltonian as follows

$$J^2 + K^2 = -\frac{1}{2}\left(\frac{mk^2}{2H} + \hbar^2\right), \tag{4.88}$$

or

$$H = -\frac{mk^2}{2}\left\{\frac{1}{2(J^2 + K^2) + \hbar^2}\right\}. \tag{4.89}$$

Since the hydrogen atom is subject to the condition $J^2 = K^2$, the eigenstates of it are then taken as the direct product of the irreducible representations of O(3) $\otimes$ O(3) as follows

$$|j, m\rangle \otimes |k, m'\rangle_{j=k} \equiv |(j, m); (k, m')\rangle_{j=k} = |(j, m); (j, m')\rangle. \tag{4.90}$$

They are also the eigenstates of Hamiltonian (why?), i.e.

$$H|(j, m); (j, m')\rangle = E|(j, m); (j, m')\rangle, \tag{4.91}$$

with the eigenvalues

$$E = -\frac{mk^2}{2}\frac{1}{4j(j + 1)\hbar^2 + \hbar^2} = -\frac{mk^2}{2}\frac{1}{(2j + 1)^2\hbar^2}, \tag{4.92}$$

or in terms of the principal quantum number $n = 2j + 1$, we reach the well known eigenenergy of the hydrogen atom,

$$E_n = -\frac{mk^2}{2\hbar^2}\frac{1}{n^2} = -\frac{mc^2}{2}\left(\frac{e^2}{\hbar c}\right)^2\frac{1}{n^2} = -\frac{mc^2}{2}\alpha^2\frac{1}{n^2},$$

where α is the fine structure constant.

4.5 Exercises

Ex 4.5.1

Show that SU(2) matrix $M(\chi \hat{n} \cdot \vec{Y}) = \exp(\chi \frac{i}{2} \hat{n} \cdot \vec{\sigma})$ corresponding to a rotation about axis $\hat{n}$ with an angle χ takes the following form,

$$M(\chi \hat{n} \cdot \vec{Y}) = \cos \frac{\chi}{2} + i\hat{n} \cdot \vec{\sigma} \sin \frac{\chi}{2}.$$

Ex 4.5.2

Taking $\sigma^i = \sigma_i$ and making use of the following properties of Pauli matrices,

(a) hermiticity $\sigma^{i\dagger} = \sigma^i$,

(b) traceless $\mathrm{Tr}\, \sigma^i = 0$,

(c) $\{\sigma^i, \sigma^j\} = 2\delta^{ij}\mathbf{I}$,

(d) $\mathrm{Tr}\, \sigma^i \sigma^j = 2\delta^{ij}$,

(e) $\det \sigma^i = -1$,

show that the rotational matrix element R^i_j is given by

$$R^i_j = \frac{1}{2}\mathrm{Tr}\, (\sigma^i M^\dagger \sigma_j M).$$

(**Hint:** by Eq. (4.36).)

Ex 4.5.3

Construct explicitly the matrix representations of the generators in two dimensions.

Ex 4.5.4

Denote the 2×2 traceless matrices A and B by arbitrary parameters as $\mathbf{A} = \vec{a} \cdot \vec{\sigma}, \mathbf{B} = \vec{b} \cdot \vec{\sigma}$. Show that $\mathbf{AB} = \vec{a} \cdot \vec{b}\mathbf{I} + i(\vec{a} \times \vec{b}) \cdot \vec{\sigma}$.

Ex 4.5.5

Show that

$$\mathbf{U}(\vec{\varphi})\vec{\sigma}\mathbf{U}^\dagger(\vec{\varphi}) = (\hat{n} \cdot \vec{\sigma})\hat{n} + \hat{n} \times \vec{\sigma} \sin\varphi - \hat{n} \times (\hat{n} \times \vec{\sigma}) \cos\varphi,$$

where $\mathbf{U}(\vec{\varphi}) = \exp(i\frac{\varphi}{2}\hat{n} \cdot \vec{\sigma})$.

Ex 4.5.6

Show that the Casimir operator of the rotational group takes the expression

$$C = \begin{cases} J_i J_i & \text{in the Cartesian basis,} \\ J_+J_- + J_-J_+ + J_0^2 & \text{in the standard form.} \end{cases}$$

Ex 4.5.7

Find the unitary matrix $\mathbf{U}$ which diagonalizes

$$J_z^{(c)} = \frac{\hbar}{i}\begin{pmatrix} 0 & -1 & 0 \\ 1 & 0 & 0 \\ 0 & 0 & 0 \end{pmatrix} \quad \text{into} \quad J_0^{(1)} = \hbar\begin{pmatrix} 1 & 0 & 0 \\ 0 & 0 & 0 \\ 0 & 0 & -1 \end{pmatrix},$$

and hence verify explicitly that

$$J_x^{(1)} = \frac{\hbar}{\sqrt{2}}\begin{pmatrix} 0 & 1 & 0 \\ 1 & 0 & 1 \\ 0 & 1 & 0 \end{pmatrix}, \quad J_y^{(1)} = \frac{\hbar}{\sqrt{2}}\begin{pmatrix} 0 & -i & 0 \\ i & 0 & -i \\ 0 & i & 0 \end{pmatrix}.$$

Ex **4.5.8**

Show that $[H, R_i] = \mathbf{O}$ for the hydrogen atom.

Ex **4.5.9**

Prove that

$$[R_i.R_j] = -\frac{2H}{mk^2}\frac{1}{\epsilon^2(1-\epsilon)^2}i\hbar\epsilon_{ijk}L_k.$$

Ex **4.5.10**

Prove that

$$J^2 + K^2 = -\frac{1}{2}\left(\frac{mk^2}{2H} + \hbar^2\right).$$

(**Hint:** use the fact that $[H, J_i] = [H, K_i] = \mathbf{O}$.)

Chapter 5

Lorentz Transformation, O(3,1)/SL(2,C) Group and the Dirac Equation

5.1 Space-time structure and Minkowski space

Space and time had been treated separately since the birth of physics as an experimental science in mediaeval time. As Newton stated in his work, The Principia, absolute space, in its own nature, without relation to anything external, remains always similar and immovable. Absolute, true, and mathematical time, of itself, and from its own nature, flows equably without relation to anything external. Physicists had been confined by the concept of the absoluteness of space and time such that the coordinate transformations from one inertial frame to another, always kept time as an invariant parameter.

All this concept of absolute space and time resulted in the formulation of Galilean relativity which has not been challenged until the end of 19th century and the beginning of the 20th century. It was the triumphal results that related the wave phenomena of light to the electromagnetic theory in Maxwell equations, that space and time were interwoven into the formulation of the theory and were treated as equals. The development in special relativity afterwards led naturally to generalize the 3-dimensional coordinate space to the 4-dimensional one by including the time-axis, an extra dimension. And so it is no longer the Euclidean vector space, but a four-dimensional Minkowski space with metric tensor defined as

$$g_{\mu\nu} = 0 \quad \text{if} \quad \mu \neq \nu \ ; \quad -g_{00} = g_{11} = g_{22} = g_{33} = 1. \qquad (5.1)$$

The zero-th axis is defined as the time axis, and $x^0 = ct$. A vector in Minkowski space, also called a **4-vector**, is denoted by $x = (x^0, x^1, x^2, x^3)$ to specify the time and position. In order to avoid using the term "norm", we define the length square of the vector as

$$x^2 = g_{\mu\nu} x^\mu x^\nu = -(x^0)^2 + (x^1)^2 + (x^2)^2 + (x^3)^2. \qquad (5.2)$$

This is a postulate in special relativity that the laws of physics are invariant under a homogeneous linear transformation on the space-time 4-vector if the transformation leaves the length square of a 4-vector invariant. This chapter is devoted to the exploration of the properties of these transformations as well as their applications in relativistic quantum wave equations.

5.1.1 Homogeneous Lorentz transformation and SO(3,1) group

Let us consider the following linear transformation in the matrix form

$$x \longmapsto x' = \Lambda x, \qquad (5.3)$$

where x and x' are column matrices, and Λ is a 4×4 square matrix.

If we denote the Minkowski metric tensor also in matrix form as

$$g_{\mu\nu} = \begin{pmatrix} -1 & 0 & 0 & 0 \\ 0 & 1 & 0 & 0 \\ 0 & 0 & 1 & 0 \\ 0 & 0 & 0 & 1 \end{pmatrix}, \qquad (5.4)$$

the length square of the 4-vector x becomes

$$x^T g x = -(x^0)^2 + (x^1)^2 + (x^2)^2 + (x^3)^2. \qquad (5.5)$$

The invariance in the square length of x due to the isotropy property of space-time restrains the Lorentz transformation matrix Λ to the following conditions

$$x'^T g x' = x^T \Lambda^T g \Lambda x = x^T g x, \qquad (5.6)$$

namely that

$$\Lambda^T g \Lambda = g \quad \text{or} \quad \Lambda^T g = g \Lambda^{-1}. \tag{5.7}$$

This is the crucial difference between the Euclidean space and the Minkowski space because the g matrix is involved in the latter case.

If $\det \Lambda \neq 0$, then the transformation forms a group, called the **homogeneous Lorentz group**. If we take the determinant on both sides of Eq. (5.7), we obtain $(\det \Lambda)^2 = 1$. We shall restrict ourselves to the case $\det \Lambda = 1$ with which the group elements can be reached continuously from the identity element. The length invariance will provide 10 equations of constraints among the matrix elements of Λ, namely

$$\Lambda^\mu_\alpha g_{\mu\nu} \Lambda^\nu_\beta = g_{\alpha\beta}. \tag{5.8}$$

Therefore there remain 6 independent parameters in matrix Λ. Furthermore for the case $\alpha = \beta = 0$, the constraint reads as

$$(\Lambda^0_0)^2 - \sum_i (\Lambda^i_0)^2 = 1, \tag{5.9}$$

which provides two choices, i.e. $\Lambda^0_0 \geqslant 1$ and $\Lambda^0_0 \leqslant 1$.

If all the group elements are continuously connected to the identity element, $\Lambda^0_0 \geqslant 1$ should be chosen, and the group is called **proper Lorentz group**, or simply the Lorentz group, or the SO(3,1) group, to which we shall confine ourselves for further investigation.

To construct the generators of the group, let us consider the **boost Lorentz transformation**, in which one spatial coordinate system $\mathcal{O}'$ is travelling with velocity in a particular direction, say along the x'^1-axis, with respect to another spatial coordinate system $\mathcal{O}$ along the direction of x^1-axis, of which both origins of the coordinate frames coincide at $x^0 = x'^0 = 0$. Then the boost matrix takes the following familiar expression:

$$x'^0 = \gamma x^0 - \beta\gamma x^1, \tag{5.10a}$$

$$x'^1 = -\beta\gamma x^0 + \gamma x^1, \tag{5.10b}$$

where $\beta = v/c$ and $\gamma = 1/\sqrt{1-\beta^2}$.

It is interesting to visualize the above boost transformation as a rotation on the (x^0, x^1) plane with an imaginary angle, or hyperbolic angle ξ defined by

$$\xi = \tanh^{-1}\beta. \tag{5.11}$$

Then the boost matrix becomes

$$\Lambda = \begin{pmatrix} \cosh\xi & -\sinh\xi & 0 & 0 \\ -\sinh\xi & \cosh\xi & 0 & 0 \\ 0 & 0 & 1 & 0 \\ 0 & 0 & 0 & 1 \end{pmatrix}. \tag{5.12}$$

The matrix representation of the generator for the boost transformation can then be easily obtained by taking the derivative of the above matrix with respect to the group parameter ξ at the origin of the parameter space, i.e.

$$B_1 = \begin{pmatrix} 0 & -1 & 0 & 0 \\ -1 & 0 & 0 & 0 \\ 0 & 0 & 0 & 0 \\ 0 & 0 & 0 & 0 \end{pmatrix}. \tag{5.13a}$$

The other two generators of the boost transformation in x^2-axis and x^3-axis can also be constructed with similar procedures. They read as:

$$B_2 = \begin{pmatrix} 0 & 0 & -1 & 0 \\ 0 & 0 & 0 & 0 \\ -1 & 0 & 0 & 0 \\ 0 & 0 & 0 & 0 \end{pmatrix}, \quad B_3 = \begin{pmatrix} 0 & 0 & 0 & -1 \\ 0 & 0 & 0 & 0 \\ 0 & 0 & 0 & 0 \\ -1 & 0 & 0 & 0 \end{pmatrix}. \tag{5.13b,c}$$

The remaining three generators of the SO(3,1) group are merely those of the spatial rotational ones, expressed as:

$$A_1 = \begin{pmatrix} 0 & 0 & 0 & 0 \\ 0 & 0 & 0 & 0 \\ 0 & 0 & 0 & 1 \\ 0 & 0 & -1 & 0 \end{pmatrix} \tag{5.14a}$$

$$\text{and } A_2 = \begin{pmatrix} 0 & 0 & 0 & 0 \\ 0 & 0 & 0 & -1 \\ 0 & 0 & 0 & 0 \\ 0 & 1 & 0 & 0 \end{pmatrix}, \quad A_3 = \begin{pmatrix} 0 & 0 & 0 & 0 \\ 0 & 0 & 1 & 0 \\ 0 & -1 & 0 & 0 \\ 0 & 0 & 0 & 0 \end{pmatrix}. \tag{5.14b,c}$$

This allows us to obtain the Lie algebra of the SO(3,1) group as follows:

$$[A_i, A_j] = -\epsilon_{ij}^k A_k, \quad [A_i, B_j] = -\epsilon_{ij}^k B_k, \tag{5.15}$$

but the commutators among the boost generators B_i take the different sign to close the algebra, i.e.

$$[B_i, B_j] = \epsilon_{ij}^k A_k. \tag{5.16}$$

The algebra of the generators A_i closes by themselves, they form a sub algebra of SO(3,1), which implies that SO(3,1) contains SO(3), the rotational transformation, as its subgroup.

As in the case of rotational transformation, the Lorentz boost matrix Λ of the relative motion along the x^1- axis can also be regained by exponentiating the matrix generator B_1, namely

$$e^{\xi B_1} = \begin{pmatrix} \cosh\xi & -\sinh\xi & 0 & 0 \\ -\sinh\xi & \cosh\xi & 0 & 0 \\ 0 & 0 & 1 & 0 \\ 0 & 0 & 0 & 1 \end{pmatrix}. \tag{5.17}$$

Therefore the group elements of the SO(3,1) can be expressed in the general forms as

$$\mathbf{\Lambda} = e^{\vec{\theta}\cdot\vec{A}+\vec{\xi}\cdot\vec{B}}. \tag{5.18}$$

Since the group parameters $\xi = (\xi^1, \xi^2, \xi^3)$ are unbounded, hence the Lorentz group, or the SO(3,1) group, is a non-compact Lie group.

The algebra of the SO(3,1) group can also be constructed by means of the usual canonical formulation, in which we denote the contravariant tensor generators as

$$M^{\mu\nu} = x^\mu \frac{\partial}{\partial x_\nu} - x^\nu \frac{\partial}{\partial x_\mu}, \tag{5.19}$$

and one will immediately confirm that

$$[M^{\mu\nu}, M^{\alpha\beta}] = -g^{\nu\beta}M^{\mu\alpha} - g^{\mu\alpha}M^{\nu\beta} + g^{\nu\alpha}M^{\mu\beta} + g^{\mu\beta}M^{\nu\alpha}. \tag{5.20}$$

It is convenient to cast the algebra in terms of A_i and B_i, which will reproduce the SO(3,1) algebra by the following identification

$$B_i = g_{ij}M^{0j}, \quad A_i = \frac{1}{2}\epsilon_{ijk}M^{jk}. \tag{5.21}$$

5.2 Irreducible representation of SO(3,1) and Lorentz spinors

As in the case of the SO(4) group, let us take the linear combinations of the generators of the SO(3,1) group and denote that

$$L_i = \frac{1}{2}\left(\frac{A_i}{i} + B_i\right), \quad \text{and} \quad R_i = \frac{1}{2}\left(\frac{A_i}{i} - B_i\right). \tag{5.22a,b}$$

Then the Lie algebra of the SO(3,1) group takes the expressions as those of the SO(3) $\otimes$ SO(3) group, namely, i.e.

$$[L_i, L_j] = i\epsilon_{ij}^k L_k, \qquad (5.23a)$$

$$[L_i, R_i] = 0, \qquad (5.23b)$$

$$[R_i, R_j] = i\epsilon_{ij}^k R_k. \qquad (5.23c)$$

Yet they are not exactly parallel to the case of the SO(4) group, because the generators L_i and R_i of the non-compact Lorentz group are not Hermitian. The immediate consequence of these properties gives rise to two options we have to face. Either we choose the finite dimensional representations, in which case the unitarity condition shall be abandoned or we choose to preserve the unitarity of the representation by accepting the infinite dimensional representations.

Let us consider the finite dimensional irreducible representation, and take the direct product $|l, m\rangle_l \otimes |r, n\rangle_r \equiv |l, m; r, n\rangle$ as the bases, where $|l, m\rangle_l$ is the eigenvector of the commuting operators L^2 and L_3 with the eigenvalues $l(l + 1)$ and m respectively, while $|r, n\rangle_r$ is the eigenvector of another pair of commuting operators R^2 and R_3 with the eigenvalues $r(r + 1)$ and n respectively, such that

$$-l \leqslant m \leqslant l, \quad -r \leqslant n \leqslant r, \quad \text{and} \quad l, r = \text{ half integers.} \qquad (5.24)$$

Consider the simplest case of one dimension in which we take $l = r = 0$, and label the representation by the symbol $(0, 0)$. The matrix representations of the generators are of one dimension, namely a 1×1 matrix with element as

$$\langle 0, 0; 0, 0|L_i|0, 0; 0, 0\rangle = \langle 0, 0; 0, 0|R_i|0, 0; 0, 0\rangle = 0, \qquad (5.25)$$

and we have the trivial representation of the group, identity.

There are two distinct 2-dimensional representations, namely $(1/2, 0)$-representation and $(0, 1/2)$-representation. We shall leave them as an exercise for the readers to calculate that the generators shall take the following forms:

$$L_i^{(\frac{1}{2},0)} = \frac{1}{2}\sigma_i, \qquad\qquad R_i^{(\frac{1}{2},0)} = 0, \qquad\qquad (5.26\text{a})$$

$$L_i^{(0,\frac{1}{2})} = 0, \qquad\qquad R_i^{(0,\frac{1}{2})} = \frac{1}{2}\sigma_i, \qquad\qquad (5.26\text{b})$$

which allow us to obtain

$$A_i^{(\frac{1}{2},0)} = \frac{i}{2}\sigma_i, \qquad\qquad B_i^{(\frac{1}{2},0)} = \frac{1}{2}\sigma_i, \qquad\qquad (5.27\text{a})$$

$$A_i^{(0,\frac{1}{2})} = \frac{i}{2}\sigma_i, \qquad\qquad B_i^{(0,\frac{1}{2})} = -\frac{1}{2}\sigma_i, \qquad\qquad (5.27\text{b})$$

and obtain finally the 2-dimensional irreducible representation of the Lorentz group as follows

$$D^{(\frac{1}{2},0)}(\vec{\theta},\vec{\xi}) = \exp\left(\frac{i}{2}\vec{\sigma}\cdot(\vec{\theta}-i\vec{\xi})\right), \qquad\qquad (5.28\text{a})$$

$$D^{(0,\frac{1}{2})}(\vec{\theta},\vec{\xi}) = \exp\left(\frac{i}{2}\vec{\sigma}\cdot(\vec{\theta}+i\vec{\xi})\right). \qquad\qquad (5.28\text{b})$$

As a quick check out of curiosity, one finds that

$$D^{(\frac{1}{2},0)\dagger}(\vec{\theta},\vec{\xi}) = \exp\left(-\frac{i}{2}\vec{\sigma}\cdot(\vec{\theta}+i\vec{\xi})\right) \neq D^{(\frac{1}{2},0)}(\vec{\theta},\vec{\xi})^{-1},$$

or

$$D^{(0,\frac{1}{2})\dagger}(\vec{\theta},\vec{\xi}) = \exp\left(-\frac{i}{2}\vec{\sigma}\cdot(\vec{\theta}-i\vec{\xi})\right) \neq D^{(0,\frac{1}{2})}(\vec{\theta},\vec{\xi})^{-1},$$

namely that unitarity is lost in the finite dimensional irreducible representation.

With the aim of investigating the **relativistic wave equation** of spin $\frac{1}{2}$ system, we shall limit ourselves by not going into further exploration of the higher dimensional representations.

Let us perform the following identifications:

$$|\frac{1}{2}, \frac{1}{2}\rangle_l \longmapsto \begin{pmatrix} 1 \\ 0 \end{pmatrix} = e_1, \quad |\frac{1}{2}, -\frac{1}{2}\rangle_l \longmapsto \begin{pmatrix} 0 \\ 1 \end{pmatrix} = e_2, \qquad (5.29a,b)$$

as the two bases of the two-dimensional complex manifold, or $\mathcal{V}_{(\frac{1}{2},0)}$-space, and construct the left-handed Lorentz spinor, or simply the **left-handed spinor** as follows

$$\psi_l(x) = \psi_l^a(x)e_a = \begin{pmatrix} \psi_l^1(x) \\ \psi_l^2(x) \end{pmatrix}, \qquad (5.30)$$

where $\psi_l^1(x)$ or $\psi_l^2(x)$ are all complex functions of space-time coordinate $x = (x^0, x^1, x^2, x^3)$. Under the Lorentz transformation, the left-handed spinor is transformed by applying the $(1/2, 0)$ matrix representation of the $SO(3,1)$ group, i.e. Eq. (5.28a) upon the spinor as follows

$$\psi_l(x) \longmapsto \psi_l'(x') = D^{(\frac{1}{2},0)}(\vec{\theta}, \vec{\xi})\psi_l(\Lambda^{-1}x'), \qquad (5.31)$$

where $x' = \Lambda^{-1}x$ implies that active transformation of the quantum states equivalent to a passive Lorentz transformation of the space-time vector. It should be emphasized that not only the components of the spinor are transformed by the $(1/2, 0)$ matrix , but that the space-time coordinates in the argument of the complex functions $\psi_l(x)$ are also simultaneously transformed accordingly.

As for the Lorentz transformation of the **right-handed spinor**, one finds that it can be expressed as

$$\psi_r(x) = \psi_r^a(x)f_a = \psi_r^1(x)\begin{pmatrix} 1 \\ 0 \end{pmatrix} + \psi_r^2(x)\begin{pmatrix} 0 \\ 1 \end{pmatrix} = \begin{pmatrix} \psi_r^1(x) \\ \psi_r^2(x) \end{pmatrix}, \qquad (5.32)$$

where $\psi_r^a(x)(a=1,2)$ are the complex functions of the space-time coordinates, and $f_a(a=1,2)$ are the bases column matrix of the right-handed spinor space. Similar to the transformation in the left-handed spinor space, the Lorentz transformation for $\psi_r(x)$ can then be written as

$$\psi_r(x) \longmapsto \psi_r'(x') = D^{(0,\frac{1}{2})}(\vec{\theta},\vec{\xi})\psi_r(\Lambda^{-1}x')). \qquad (5.33)$$

5.3 SL(2,C) group and the Lorentz transformation

It is interesting to observe that both matrices $D^{(\frac{1}{2},0)}(\vec{\theta},\vec{\xi})$ and $D^{(0,\frac{1}{2})}(\vec{\theta},\vec{\xi})$ play more roles than just performing the Lorentz transformation upon the spinor wave functions. Let us consider the linear transformation on a two-dimensional complex vector space, in which a vector ξ is acted upon by a 2×2 matrix L with complex matrix elements as follows

$$\xi' = \begin{pmatrix} \xi'^1 \\ \xi'^2 \end{pmatrix} = L\begin{pmatrix} \xi^1 \\ \xi^2 \end{pmatrix} = \begin{pmatrix} a & b \\ c & d \end{pmatrix}\begin{pmatrix} \xi^1 \\ \xi^2 \end{pmatrix}. \qquad (5.34)$$

Let us impose a condition such that

$$\det L = \begin{vmatrix} a & b \\ c & d \end{vmatrix} = ab - bc = 1, \qquad (5.35)$$

which reduces the 8 parameters to 6 independent ones. Matrices with the properties of such specification form a group of SL(2,C). If we express the SL(2,C) matrix by exponentiating an arbitrary 2×2 matrix A, namely

$$L = e^A, \qquad (5.36)$$

then we have the following proposition

> **Proposition 1.**
>
> If a matrix L can be expressed as $L = e^{A}$, then
> $$\det L = e^{\text{Tr } A}. \qquad (5.37)$$

Proof of the proposition is straightforward: we find the eigenvectors of the matrix A as the new bases that form the matrix S, which diagonalizes the matrix A, i.e.

$$A' = S^{-1}AS. \qquad (5.38)$$

If we insert the matrix L between S^{-1} and S, we obtain the following relation

$$S^{-1}LS = e^{S^{-1}AS} = e^{A'} = \exp\begin{pmatrix} a^1 & 0 \\ 0 & a^2 \end{pmatrix} = \begin{pmatrix} e^{a^1} & 0 \\ 0 & e^{a^2} \end{pmatrix}. \qquad (5.39)$$

Therefore we conclude that

$$\det L = \det (S^{-1}LS) = e^{a^1 + a^2} = e^{\text{Tr } A'} = e^{\text{Tr } S^{-1}AS} = e^{\text{Tr } A}. \qquad (5.40)$$

Let us go back to the two dimensional irreducible representations of the Lorentz group, the matrix $D^{(\frac{1}{2},0)}(\vec{\theta},\vec{\xi})$ and matrix $D^{(0,\frac{1}{2})}(\vec{\theta},\vec{\xi})$. They are in fact the SL(2,C) matrices because

$$\det D^{(\frac{1}{2},0)}(\vec{\theta},\vec{\xi}) = \det e^{\frac{i}{2}\vec{\sigma}\cdot(\vec{\theta}-i\vec{\xi})} = e^{\text{Tr } \frac{i}{2}\vec{\sigma}\cdot(\vec{\theta}-i\vec{\xi})} = 1, \qquad (5.41)$$

and similarly

$$\det D^{(0,\frac{1}{2})}(\vec{\theta},\vec{\xi}) = \det e^{\frac{i}{2}\vec{\sigma}\cdot(\vec{\theta}+i\vec{\xi})} = e^{\text{Tr } \frac{i}{2}\vec{\sigma}\cdot(\vec{\theta}+i\vec{\xi})} = 1. \qquad (5.42)$$

The 6 group parameters are designated as $\vec{\theta} = (\theta^1,\theta^2,\theta^3)$ and $\vec{\xi} = (\xi^1,\xi^2,\xi^3)$, which enable us to formulate the generators of the SL(2,C) group in the following matrix representations, i.e.

$$A_i = \frac{i}{2}\sigma_i, B_i = \frac{1}{2}\sigma_i \quad \text{or} \quad A_i = \frac{i}{2}\sigma_i, B_i = -\frac{1}{2}\sigma_i, \qquad (5.43\text{a,b})$$

with the same Lie algebra of the SO(3,1) group, i.e.

$$[A_i, A_j] = -\epsilon_{ij}^k A_k, \qquad (5.44\text{a})$$

$$[A_i, B_j] = -\epsilon_{ij}^k B_k, \qquad (5.44\text{b})$$

$$[B_i, B_j] = \epsilon_{ij}^k A_k. \qquad (5.44\text{c})$$

We conclude therefore that SL(2,C) is an isomorphism onto the SO(3,1) group. To demonstrate the relation between the two groups with respect to Lorentz transformation, we introduce the space-time matrix

$$\boldsymbol{X} = x^\mu \boldsymbol{\sigma}_\mu = \begin{pmatrix} -x^0 + x^3 & x^1 - ix^2 \\ x^1 + ix^2 & -x^0 - x^3 \end{pmatrix}, \qquad (5.45)$$

where $\sigma^0 = -\sigma_0 = \boldsymbol{I}$, a unit matrix, and $\boldsymbol{\sigma}_i$ are the Pauli matrices. The length of the space-time vector is related the determinant of the matrix, i.e.

$$\det \boldsymbol{X} = (x^0)^2 - \vec{x}^2. \qquad (5.46)$$

Therefore the transformation of $\boldsymbol{X}$ into $\boldsymbol{X}'$ by multiplying SL(2,C) matrices, $D^{(\frac{1}{2},0)}(\vec{\theta}, \vec{\xi})$ and $D^{(\frac{1}{2},0)\dagger}(\vec{\theta}, \vec{\xi})$ on both sides of $\boldsymbol{X}$ in the following construction

$$\boldsymbol{X}' = D^{(\frac{1}{2},0)}(\vec{\theta}, \vec{\xi}) \boldsymbol{X} D^{(\frac{1}{2},0)\dagger}(\vec{\theta}, \vec{\xi}), \qquad (5.47)$$

will leave the length of the vector $x = (x^0, x^1, x^2, x^3)$ invariant because

$$\det \boldsymbol{X}' = \det \boldsymbol{X}. \qquad (5.48)$$

Take a specific example of a Lorentz boost transformation along the x^3-axis, in which $\vec{\theta} = (0, 0, 0)$, and $\vec{\xi} = (0, 0, \xi)$, then we obtain

$$\boldsymbol{X}' = e^{\frac{1}{2}\sigma_3\xi}\boldsymbol{X}e^{\frac{1}{2}\sigma_3\xi} = \begin{pmatrix} e^{\frac{1}{2}\xi} & 0 \\ 0 & e^{-\frac{1}{2}\xi} \end{pmatrix} \boldsymbol{X} \begin{pmatrix} e^{\frac{1}{2}\xi} & 0 \\ 0 & e^{-\frac{1}{2}\xi} \end{pmatrix}. \tag{5.49}$$

Therefore we recover the usual Lorentz transformation by boosting the coordinate frame along the x^3-axis if we compare the matrix elements on both sides of the last equation and obtain the following relations:

$$x'^0 = \cosh\xi x^0 - \sinh\xi x^3, \tag{5.50a}$$

$$x'^1 = x^1, \tag{5.50b}$$

$$x'^2 = x^2, \tag{5.50c}$$

$$x'^3 = -\sinh\xi x^0 + \cosh\xi x^3. \tag{5.50d}$$

5.4 Chiral transformation and spinor algebra

As we have mentioned before, physics involves two distinct types of transformations, the continuous transformations and the discrete ones. Here in this section, we shall introduce another discrete symmetry transformation. In addition to parity, charge conjugation or time reversal transformations in quantum mechanics, the chiral transformation is among the least discussed topics. The word "chirality" was derived from the Greek $\chi\varepsilon\iota\rho$, which means hand. Henceforth chirality is generally recognized as synonym to handedness in the geometric sense. It is a discrete transformation from the right-handed irreducible representation to the left-handed one of the Lorentz group, and vice versa. In contrast to the usual parity transformation, an antilinear and antiunitary operator is required in the chiral transformation in order to obtain the self consistent theory, such that the Lorentz transformation can equivalently be performed in the left-handed frame of reference as well as the right-handed one. It also provides the interrelation between the left-handed spinors and the right-handed spinors.

Let us denote the chiral operator $\mathcal{K}$, an antilinear operator and for each vector ψ, there exists a vector $a\psi + b\varphi$ such that

$$\mathcal{K}(a\psi + b\varphi) = a^*\mathcal{K}\psi + b^*\mathcal{K}\varphi, \qquad (5.51)$$

where a and b are complex numbers. The antiunitarity of $\mathcal{K}$ is defined for each pair of vectors ψ and φ, such that the inner product takes the following relation

$$(\mathcal{K}\psi, \mathcal{K}\varphi) = (\varphi, \psi) = (\psi, \varphi)^*. \qquad (5.52)$$

The chiral transformation is neither the parity transformation on the spatial coordinate system nor is it the time reversal transformation on the temporal axis in Minkowski space. The six generators of the Lorentz group therefore are invariant under the chiral transformation, namely the rotational generators as well as the Lorentz boost generators remain unchanged under $\mathcal{K}$, i.e.

$$\mathcal{K}A_i\mathcal{K}^{-1} = A_i, \quad \mathcal{K}B_i\mathcal{K}^{-1} = B_i. \qquad (5.53)$$

Yet the new set of the group generators L_i and R_i are transformed according to the following relations

$$\mathcal{K}L_i\mathcal{K}^{-1} = \frac{1}{2}\mathcal{K}\left(\frac{A_i}{i} + B_i\right)\mathcal{K}^{-1} = -R_i, \quad \mathcal{K}R_i\mathcal{K}^{-1} = -L_i, \quad (5.54\text{a,b})$$

because of the antilinearity property of operator $\mathcal{K}$. Since chiral transformation on L^2 becomes R^2 and reciprocally it holds for R^2, namely the chiral transformation will interchange operator L^2 and operator R^2, therefore we have the following proposition:

Proposition 2.

The vector $\mathcal{K}L_{jm}$ is the eigenvector of R^2 and R_3 with the eigenvalues $j(j+1)$ and $-m$ respectively. While the vector $\mathcal{K}R_{kn}$ is the eigenvector of L^2 and L_3 with the eigenvalues $k(k+1)$ and $-n$ respectively.

The proof goes as follows: let us use the brief notations L_{jm}, R_{kn} instead of $|j,m\rangle_l$ and $|k,n\rangle_r$ respectively, and consider $R^2 \mathcal{K} L_{jm} = \mathcal{K} L^2 L_{jm} = j(j+1)\mathcal{K} L_{jm}$, and $R_3 \mathcal{K} L_{jm} = -\mathcal{K} L_3 L_{jm} = -m\mathcal{K} L_{jm}$ which implies that $\mathcal{K} L_{jm} = \gamma(m) R_{j,-m}$. Similarly that $\mathcal{K} R_{kn} = \delta(n) \times L_{k,-n}$.

The antiunitarity of the operator $\mathcal{K}$ allows us to find the coefficients $\gamma(m)$ and $\delta(n)$ as follows

$$(\mathcal{K} L_{jm}, \mathcal{K} L_{jm'}) = \gamma^*(m)\gamma(m')(R_{j,-m}, R_{j,-m'}) = (L_{jm'}, L_{jm}), \quad (5.55)$$

or $\gamma^*(m)\gamma(m')\delta_{-m,-m'} = \delta_{m'm}$, that leads to $|\gamma(m)|^2 = 1$.

The same argument can be applied to the right-handed case to obtain $|\delta(n)|^2 = 1$. It is of interest to observe that the chiral transformation will also make the connection for the irreducible representation of the left-handed ones to the right-handed ones. Let us take the spinor representation as an example by considering the matrix element $D_{mm'}^{(\frac{1}{2},0)}$ of the left-handed Lorentz transformation,

$$D_{mm'}^{(\frac{1}{2},0)} = (L_{\frac{1}{2},m}, e^{\vec{\theta}\cdot\vec{A}+\vec{\xi}\cdot\vec{B}} L_{\frac{1}{2},m'}) = (\mathcal{K} e^{\vec{\theta}\cdot\vec{A}+\vec{\xi}\cdot\vec{B}} L_{\frac{1}{2},m'}, \mathcal{K} L_{\frac{1}{2},m})$$

$$= \gamma^*(m')\gamma(m)(e^{\vec{\theta}\cdot\vec{A}+\vec{\xi}\cdot\vec{B}} R_{\frac{1}{2},-m'}, R_{\frac{1}{2},-m}) = \gamma(m) D_{-m,-m'}^{(0,\frac{1}{2})*} \gamma^*(m'),$$

or it can be cast into the matrix form as follows

$$D^{(\frac{1}{2},0)} = \begin{pmatrix} 0 & \gamma(\frac{1}{2}) \\ \gamma(-\frac{1}{2}) & 0 \end{pmatrix} D^{(0,\frac{1}{2})*} \begin{pmatrix} 0 & \gamma^*(-\frac{1}{2}) \\ \gamma^*(\frac{1}{2}) & 0 \end{pmatrix}. \quad (5.56)$$

Similarly the right-handed Lorentz matrix is related to the left-handed one by

$$D^{(0,\frac{1}{2})} = \begin{pmatrix} 0 & \delta(\frac{1}{2}) \\ \delta(-\frac{1}{2}) & 0 \end{pmatrix} D^{(\frac{1}{2},0)*} \begin{pmatrix} 0 & \delta^*(-\frac{1}{2}) \\ \delta^*(\frac{1}{2}) & 0 \end{pmatrix}. \quad (5.57)$$

By choosing the phase factors $\gamma(\frac{1}{2}) = \delta(\frac{1}{2}) = -\gamma(-\frac{1}{2}) = -\delta(-\frac{1}{2}) = 1$,

and defining a 2×2 matrix ϵ as $\begin{pmatrix} 0 & 1 \\ -1 & 0 \end{pmatrix}$, one can easily prove for $i = 1, 2$ and 3 that

$$\epsilon \sigma_i^* \epsilon^{-1} = \begin{pmatrix} 0 & 1 \\ -1 & 0 \end{pmatrix} \sigma_i^* \begin{pmatrix} 0 & -1 \\ 1 & 0 \end{pmatrix} = -\sigma_i, \qquad (5.58)$$

which allows one to reciprocate the following relation:

$$D^{(0,\frac{1}{2})} = \epsilon D^{(\frac{1}{2},0)*} \epsilon^{-1}, \qquad (5.59a)$$

$$D^{(\frac{1}{2},0)} = \epsilon D^{0,(\frac{1}{2})*} \epsilon^{-1}, \qquad (5.59b)$$

We are now in the position to investigate further the connection between the SL(2,C) transformation and the Lorentz transformation with spinor algebra. Let us construct a dual space $\mathcal{V}_{(\frac{i}{2},0)}$ to the spinor space $\mathcal{V}_{(\frac{1}{2},0)}$, then $\mathcal{V}_{(\frac{i}{2},0)}$ is called the **co-left-handed spinor** space, a 2-dimensional complex vector space with the basis $\dot{e}^a (a = 1, 2)$ which can be identified with the row matrix as follows

$$\dot{e}^1 = (1, 0), \quad \dot{e}^2 = (0, 1), \qquad (5.60a,b)$$

where a new set of notations are introduced in the subscript indices in order to differentiate the co-left-handed spinor to the right-handed one.

A spinor in $\mathcal{V}_{(\frac{i}{2},0)}$ is then expressed by

$$\dot{\psi} = \dot{\psi}_a \dot{e}^a = (\dot{\psi}_1, \dot{\psi}_2), \qquad (5.61)$$

where $\dot{\psi}$ is related to the components of the left-handed spinor $\psi = (\psi^1, \psi^2)^T$ by the following equation

$$\dot{\psi}_a = \psi^b \epsilon_{ba}, \qquad (5.62)$$

or in matrix notation as

$$\dot{\psi} = \psi^T \epsilon^T = (\dot{\psi}_1, \dot{\psi}_1) = (\psi^1, \psi^2) \epsilon^T = (\psi^1, \psi^2) \epsilon^{-1}. \qquad (5.63)$$

The Lorentz transformation on the co-left-handed spinor can be found to be

$$\dot{\psi} \xmapsto{L.T.} \dot{\psi}' = \psi'^T \epsilon^{-1} = \psi^T D^{T(\frac{1}{2},0)} \epsilon^{-1}$$

$$= \dot{\psi}(\epsilon D^{(\frac{1}{2},0)*} \epsilon^{-1})^\dagger = \dot{\psi} D^{(0,\frac{1}{2})\dagger}.$$

Similarly if we construct a **co-right-handed spinor** $\dot{\varphi} = (\dot{\varphi}_1, \dot{\varphi}_2)$, the dual vector to the right-handed spinor $\varphi = \begin{pmatrix} \varphi_1 \\ \varphi_2 \end{pmatrix}$ is defined as follows

$$\dot{\varphi} = \varphi^T \epsilon^T. \tag{5.64}$$

Then the Lorentz transformation on $\dot{\varphi} = (\dot{\varphi}_1, \dot{\varphi}_2)$ can be expressed as

$$\dot{\varphi} \xmapsto{L.T.} \dot{\varphi}' = \varphi'^T \epsilon^T = \dot{\varphi} D^{(\frac{1}{2},0)\dagger}. \tag{5.65}$$

Let us proceed to consider the irreducible representation denoted by $(\frac{1}{2}, \frac{\dot{1}}{2})$. It is a 4-dimensional vector space spanned by the direct product of two spinor spaces, namely the left-handed spinor spaces and the co-right-handed spinor space, with the bases given by $e_a \dot{f}^b$. A general element in $(\frac{1}{2}, \frac{\dot{1}}{2})$-space can be expressed as $u^a_{\dot{b}} e_a \dot{f}^b$, or if cast into the matrix formulation, it can be written as

$$U^{(\frac{1}{2},\frac{\dot{1}}{2})} = \begin{pmatrix} u^1_{\dot{1}} & u^1_{\dot{2}} \\ u^2_{\dot{1}} & u^2_{\dot{2}} \end{pmatrix}. \tag{5.66}$$

A similar argument leads to express the $(\frac{\dot{1}}{2}, \frac{1}{2})$ representation as follows

$$U^{(\frac{\dot{1}}{2},\frac{1}{2})} = \begin{pmatrix} u^{\dot{1}}_1 & u^{\dot{2}}_1 \\ u^{\dot{1}}_2 & u^{\dot{2}}_2 \end{pmatrix}. \tag{5.67}$$

It is obvious that Lorentz transformation on vectors in $(\frac{1}{2}, \frac{\dot{1}}{2})$-space and in $(\frac{\dot{1}}{2}, \frac{1}{2})$- space, takes respectively the following expressions:

$$U^{(\frac{1}{2},\frac{1}{2})} \xrightarrow{L.T.} U'^{(\frac{1}{2},\frac{1}{2})} = D^{(\frac{1}{2},0)}UD^{(\frac{1}{2},0)\dagger}, \qquad (5.68a)$$

$$U^{(\frac{1}{2},\frac{1}{2})} \xrightarrow{L.T.} U'^{(\frac{1}{2},\frac{1}{2})} = D^{(0,\frac{1}{2})}UD^{(0,\frac{1}{2})\dagger}. \qquad (5.68b)$$

This transformation property immediately convinces us that the space-time matrix

$$\boldsymbol{X} = x^\mu \sigma_\mu = \begin{pmatrix} -x^0 + x^3 & x^1 - ix^2 \\ x^1 + ix^2 & -x^0 - x^3 \end{pmatrix}, \qquad (5.69)$$

which we have constructed in Section 5.3 is an operator belonging to $(\frac{1}{2}, \frac{1}{2})$-representation.

5.5 Lorentz spinors and the Dirac equation

As we have discussed in Section 5.1, time and space are regarded as a four vector in the Minkowski space. In fact, the energy and the momentum of a mass point, can also be treated as a Minkowski four vector in relativistic dynamics. Let us define the world line as the trajectory of a mass point in the four-dimensional Minkowski space in which the space-time coordinates are functions of a parameter τ, called the **proper time**, a Lorentz invariant scalar whose differential is defined as follows

$$(d\tau)^2 = -\frac{1}{c^2} g_{\mu\nu} dx^\mu dx^\nu. \qquad (5.70)$$

Consider the Lorentz frame which is boosted instantaneously with the same velocity as that of the moving mass point. Then the time measured in this frame where the mass point is instantaneously at rest, namely that $d\vec{x}' = 0$, is in fact the proper time because of the invariant property,

$$d\tau = d\tau' = \left(-\frac{1}{c^2} g_{\mu\nu} dx'^\mu dx'^\nu \right)^{\frac{1}{2}} = dt'. \qquad (5.71)$$

Yet this invariant quantity when measured in the initial reference frame, namely $\mathcal{O}$-system, is given by

$$d\tau = \left(-\frac{1}{c^2}g_{\mu\nu}dx^\mu dx^\nu\right)^{\frac{1}{2}} = \sqrt{1 - \frac{v^2}{c^2}}dt. \qquad (5.72)$$

We will then introduce a velocity 4-vector $u^\mu(\tau)$ defined as $u^\mu(\tau) = dx^\mu/d\tau$ which is related to the energy-momentum 4-vector p^μ by the following equation

$$p^\mu = m_0 u^\mu. \qquad (5.73)$$

The energy-momentum 4-vector for a mass point moving with velocity $\vec{v}$ can be expressed in the reference frame as $p^\mu = (p^0, \vec{p})$, which when evaluated in the usual relativistic form becomes:

$$p^0 = m_0\frac{dx^0}{d\tau} = \frac{m_0 c}{\sqrt{1 - \beta^2}}, \qquad (5.74a)$$

$$\vec{p} = m_0\frac{d\vec{x}}{d\tau} = \frac{m_0 c\vec{\beta}}{\sqrt{1 - \beta^2}}. \qquad (5.74b)$$

One can easily verify that the scalar product, or the length of the 4-vector p^μ is characterized by the rest mass of the moving particle, an invariant under the Lorentz transformation, namely

$$p_\mu p^\mu = -m_0^2 c^2. \qquad (5.75)$$

Let us construct a 2×2 energy-momentum matrix as follows

$$\boldsymbol{P} = p^\mu \boldsymbol{\sigma}_\mu = \begin{pmatrix} -p^0 + p^3 & p^1 - ip^2 \\ p^1 + ip^2 & -p^0 - p^3 \end{pmatrix}. \qquad (5.76)$$

It transforms as the $(\frac{1}{2}, \frac{1}{2})$ irreducible representation of the Lorentz group, i.e.

$$\boldsymbol{P} \overset{L.T.}{\longmapsto} \boldsymbol{P}' = D^{(\frac{1}{2},0)}(\vec{\theta}, \vec{\xi}) \boldsymbol{P} D^{(\frac{1}{2},0)\dagger}(\vec{\theta}, \vec{\xi}). \tag{5.77}$$

As a particular case, let us consider a mass point at rest at the origin of a frame $\mathcal{O}$, and a frame $\mathcal{O}'$ which is travelling along x^3- axis with the relative velocity v. Then the 4-momentum in the $\mathcal{O}'$ system can be expressed as

$$\boldsymbol{P}' = e^{\frac{1}{2}\sigma_3\xi} \boldsymbol{P} e^{\frac{1}{2}\sigma_3\xi^\dagger}, \tag{5.78}$$

or in matrix notation,

$$\boldsymbol{P}' = \begin{pmatrix} e^{\frac{1}{2}\xi} & 0 \\ 0 & e^{-\frac{1}{2}\xi} \end{pmatrix} \begin{pmatrix} -m_0 c & 0 \\ 0 & -m_0 c \end{pmatrix} \begin{pmatrix} e^{\frac{1}{2}\xi} & 0 \\ 0 & e^{-\frac{1}{2}\xi} \end{pmatrix}$$

$$= \begin{pmatrix} -m_0 c e^{\xi} & 0 \\ 0 & -m_0 c e^{-\xi} \end{pmatrix}. \tag{5.79}$$

We expect that the energy and the momentum of the mass point, when measured in the $\mathcal{O}'$-system, will be respectively given as follows:

$$p'^0 = m_0 c \cosh\xi = \frac{m_0 c}{\sqrt{1-\beta^2}}, \tag{5.80a}$$

$$p'^3 = -m_0 c \sinh\xi = \frac{-m_0 v}{\sqrt{1-\beta^2}}. \tag{5.80b}$$

Let us consider the chiral conjugate (abbreviated as c.c.) of the space-time matrix $\boldsymbol{X}$ defined as follows

$$\boldsymbol{X} \overset{c.c.}{\longmapsto} \boldsymbol{X}_c = \epsilon \boldsymbol{X}^* \epsilon^{-1}. \tag{5.81}$$

The Lorentz transformation on $\boldsymbol{X}_c$ can be evaluated by the transformation property of $\boldsymbol{X}$, namely that

$$\boldsymbol{X}_c' = \epsilon \boldsymbol{X}'^* \epsilon^{-1} = \epsilon (D^{(\frac{1}{2},0)} \boldsymbol{X} D^{(\frac{1}{2},0)\dagger})^* \epsilon^{-1} = D^{(0,\frac{1}{2})} \boldsymbol{X}_c D^{(0,\frac{1}{2})\dagger}. \tag{5.82}$$

One recognizes that $\boldsymbol{X}$ and $\boldsymbol{X}_c$ belong respectively to the $(\frac{1}{2}, \frac{1}{2})$ and the

$(\frac{1}{2}, \frac{1}{2})$ irreducible representations. Yet as far as Lorentz transformation on the space-time coordinate is concerned, both matrices $D^{(\frac{1}{2},0)}$ and $D^{(0,\frac{1}{2})}$ play equivalent roles but result in the very same Lorentz transformation, namely we have two different ways to perform the same Lorentz transformation by means of employing two different SL(2,C) matrices, i.e. $D^{(\frac{1}{2},0)}$ and $D^{(0,\frac{1}{2})}$.

The space-time operator $\boldsymbol{X}$ is transformed as $(\frac{1}{2}, \frac{1}{2})$ representation under the Lorentz group. If this tensor operator acts upon a vector in the right-handed spinor space, i.e. $\boldsymbol{X}\varphi$, it turns out that the new transformed vector is a left-handed spinor , because the contraction of the second indices of the $(\frac{1}{2}, \frac{1}{2})$-tensor with the indices of the $(0, \frac{1}{2})$-tensor results in the $(\frac{1}{2}, 0)$-tensor, namely, a left-handed spinor. We shall summarize these properties in the following proposition:

Proposition 3.

An operator of the $(\frac{1}{2}, \frac{1}{2})$-representation that acts upon a vector of the $(0, \frac{1}{2})$-representation yields a vector of $(\frac{1}{2}, 0)$-representation. Conversely an operator of the $(\frac{1}{2}, \frac{1}{2})$-representation that acts upon a vector of the $(\frac{1}{2}, 0)$-representation will yield a vector of the $(0, \frac{1}{2})$-representation.

Let $\boldsymbol{A}$ be an operator of the $(\frac{1}{2}, \frac{1}{2})$-representation, and ξ a vector of $(0, \frac{1}{2})$-representation, then if we denote η as follows

$$\eta = \boldsymbol{A}\xi, \tag{5.83}$$

then the Lorentz transformation on η becomes

$$\eta \xmapsto{L.T.} \eta' = \boldsymbol{A}'\xi' = D^{(\frac{1}{2},0)} \boldsymbol{A} D^{(\frac{1}{2},0)\dagger} D^{(0,\frac{1}{2})}\xi = D^{(\frac{1}{2},0)}\eta, \tag{5.84}$$

which implies that η is a vector of the $(\frac{1}{2}, 0)$-representation. Similarly if we denote $\boldsymbol{A}_c$, an operator of the $(\frac{1}{2}, \frac{1}{2})$-representation, then $\xi = \boldsymbol{A}_c\eta$ has the following transformation property,

$$\xi \xmapsto{L.T.} \xi' = \boldsymbol{A}'_c \eta' = D^{(0,\frac{1}{2})} \boldsymbol{A}_c D^{(0,\frac{1}{2})\dagger} D^{(\frac{1}{2},0)} \xi = D^{(0,\frac{1}{2})} \xi, \qquad (5.85)$$

which states that ξ is the vector of the $(0, \frac{1}{2})$-representation.

The above proposition allows us to conclude that $\boldsymbol{P}_c \psi_l$ and $\boldsymbol{P} \psi_r$ are the right-handed spinor and the left-handed spinor respectively. If we represent the quantum state of a free mass point with spin equal to $1/2$ by the spinors, then we have

$$\boldsymbol{P}_c \psi_l = m_0 c \psi_r, \qquad (5.86a)$$

$$\boldsymbol{P} \psi_r = m_0 c \psi_l, \qquad (5.86b)$$

where we introduce the proportional constant $m_0 c$ in order to obtain the same physical dimension on both sides of the equation.

If we take the **direct sum** of the right-handed spinor and the left-handed spinor to form a 4-component spinor, called **Dirac spinor**, the above two equations can be simplified in the following expression

$$\begin{pmatrix} 0 & \boldsymbol{P}_c \\ \boldsymbol{P} & 0 \end{pmatrix} \begin{pmatrix} \psi_r \\ \psi_l \end{pmatrix} = m_0 c \begin{pmatrix} \psi_r \\ \psi_l \end{pmatrix}, \qquad (5.87)$$

or

$$\begin{pmatrix} 0 & \boldsymbol{P}_c \\ \boldsymbol{P} & 0 \end{pmatrix} \psi_d(x) = m_0 c \psi_d(x), \qquad (5.88)$$

where $\psi_d(x) = \psi_r(x) \oplus \psi_l(x)$ is the Dirac spinor.

The Dirac equation is usually expressed in the q-representation by identifying that

$$P_\mu = \frac{\hbar}{i} \frac{\partial}{\partial x^\mu} = \frac{\hbar}{i} \partial_\mu. \qquad (5.89)$$

The Dirac equation then reads as

$$\left[\begin{pmatrix} 0 & i\sigma_c^\mu \\ i\sigma^\mu & 0 \end{pmatrix} \partial_\mu + \frac{m_0 c}{\hbar} \right] \psi_d(x) = 0, \tag{5.90}$$

or often expressed in covariant form as

$$\left(i\gamma^\mu \partial_\mu + \frac{m_0 c}{\hbar} \right) \psi_d(x) = 0, \tag{5.91}$$

where the 4×4 matrices γ^μ are called **Dirac matrices** with the explicit expression given as follows:

$$\gamma^\mu = \begin{pmatrix} 0 & \sigma_c^\mu \\ \sigma^\mu & 0 \end{pmatrix} \quad \text{or} \quad \gamma^0 = \begin{pmatrix} 0 & \mathbf{I} \\ \mathbf{I} & 0 \end{pmatrix}, \quad \gamma^i = \begin{pmatrix} 0 & -\sigma^i \\ \sigma^i & 0 \end{pmatrix}. \tag{5.92}$$

By making use of the following property of the anti-commutator in γ^μ, namely that

$$\{\gamma^\mu, \gamma^\nu\} = \gamma^\mu \gamma^\nu + \gamma^\nu \gamma^\mu = -2g^{\mu\nu} \mathbf{I}, \tag{5.93}$$

we are able to convert the Dirac equation which is of the first order derivative in space and time, into the second order differential equation of the Klein-Gordon one. Applying the factor $i\gamma^\nu \partial_\nu$ once more upon the Dirac equation, we then obtain the following relation

$$\left(i\gamma^\nu \partial_\nu i\gamma^\mu \partial_\mu + \frac{m_0 c}{\hbar} i\gamma^\nu \partial_\nu \right) \psi_d(x) = \left[g^{\mu\nu} \partial_\mu \partial_\nu - \left(\frac{m_0 c}{\hbar} \right)^2 \right] \psi_d(x) = 0,$$

or

$$\left[\partial_\mu \partial^\mu - \left(\frac{m_0 c}{\hbar} \right)^2 \right] \psi_d(x) = 0, \tag{5.94}$$

which implies that each component in the Dirac spinor satisfies the **Klein-Gordon equation**.

The covariant formulation of the Dirac equation does not imply that $\gamma^\mu \partial_\mu$ is Lorentz invariant, namely that $i\gamma^\mu \partial_\mu$ in the $\mathcal{O}$-Lorentz frame will not take the expression $i\gamma'^\mu \partial'_\mu$ in the $\mathcal{O}'$-Lorentz frame. Once the Dirac matrices are chosen in one Lorentz frame, they will in fact be valid in all of the Lorentz frames. The universality of the γ^μ matrices is the unique feature of the Dirac equation. What has to be modified in the Dirac equation under Lorentz transformation are the space-time coordinates as well as the Dirac spinor components, namely the Lorentz transformed Dirac equation in the $\mathcal{O}'$-system will take the following form

$$\left(i\gamma^\mu \partial'_\mu + \frac{m_0 c}{\hbar} \right) \psi'_d(x') = 0, \tag{5.95}$$

with exactly the same γ^μ matrices appearing as in the $\mathcal{O}$-system. We shall prove the statement in the following proposition.

Proposition 4.

The gamma matrices γ^μ are universal in all Lorentz frame, namely the Dirac equation in another Lorentz frame, i.e. the $\mathcal{O}'$-system always takes the same gamma matrices γ^μ used in $\mathcal{O}$-system. The equation in $\mathcal{O}'$-system is expressed as

$$\left(i\gamma^\mu \partial'_\mu + \frac{m_0 c}{\hbar} \right) \psi'_d(x') = 0.$$

The proof of the statement goes as follows: let us perform the Lorentz transformation of the Dirac equation in $\mathcal{O}$-system by multiplying the following 4×4 matrix $D(\vec{\theta}, \vec{\xi})$, the direct sum of the irreducible representation $D^{(0,\frac{1}{2})}(\vec{\theta}, \vec{\xi})$ and $D^{(\frac{1}{2},0)}(\vec{\theta}, \vec{\xi})$, i.e.

$$D(\vec{\theta}, \vec{\xi}) = D^{(0,\frac{1}{2})}(\vec{\theta}, \vec{\xi}) \oplus D^{(\frac{1}{2},0)}(\vec{\theta}, \vec{\xi})$$

$$D(\vec{\theta}, \vec{\xi}) \left(i\gamma^\mu \partial_\mu + \frac{m_0 c}{\hbar} \right) \psi_d(x) = 0.$$

If we insert an identity $\mathbf{I} = D^{-1}D$ right after the γ^μ matrix, then the above equation reads as

$$\left(iD\gamma^\mu D^{-1}\partial_\mu + \frac{m_0 c}{\hbar} \right) D\psi_d(x) = 0. \tag{5.96}$$

We shall leave it as an exercise for the readers to show that

$$D\gamma^\mu D^{-1} = \Lambda^\mu_\nu \gamma^\nu, \tag{5.97}$$

which allows us to express the Dirac equation in $\mathcal{O}'$-system as

$$\left(i\gamma^\mu \partial'_\mu + \frac{m_0 c}{\hbar} \right) \psi'_d(x') = 0,$$

by identifying $\psi'_d(x') \equiv D(\vec{\theta}, \vec{\xi})\psi_d(x) = D(\vec{\theta}, \vec{\xi})\psi_d(\Lambda^{-1}x')$.

The representations of the Dirac matrices γ^μ are not unique. During the development of quantum theory, the formulation of γ^μ matrices differed from one school to another. The γ^μ matrices we adopt in Eq. (5.92) are called the **Weyl representation.**

Let us define a new set of gamma matrices by means of similarity transformation with the matrix S given as follows

$$S = \frac{1}{\sqrt{2}} \begin{pmatrix} \mathbf{I} & \mathbf{I} \\ -\mathbf{I} & \mathbf{I} \end{pmatrix}, \tag{5.98}$$

and denote the transformed matrices by $\tilde{\gamma}_\mu$. Then the new set of Dirac matrices reads as follows

$$\tilde{\gamma}^\mu = S\gamma^\mu S^{-1}, \tag{5.99}$$

or explicitly:

$$\tilde{\gamma}^0 = \begin{pmatrix} \mathbf{I} & 0 \\ 0 & \mathbf{I} \end{pmatrix}, \quad \tilde{\gamma}^i = \begin{pmatrix} 0 & -\boldsymbol{\sigma}^i \\ \boldsymbol{\sigma}^i & 0 \end{pmatrix}, \tag{5.100}$$

which are commonly referred to as the **standard representation** of
the gamma matrices or the **Dirac-Pauli representation** as it is often
called.

The similarity transformation of the gamma matrices reshuffles the
components of a Dirac spinor in order to maintain exactly the same form
of the wave equation. Let us multiply the whole equation of Eq. (5.91)
by the matrix S, i.e.

$$S \left(i\gamma^\mu \partial_\mu + \frac{m_0 c}{\hbar} \right) \psi_d(x) = 0.$$

If we insert an identity $\mathbf{I} = S^{-1}S$ right after γ^μ in the last equation,
and denote the new Dirac spinor by $\tilde{\psi}_d(x)$ and $\tilde{\gamma}^\mu = S\gamma^\mu S^{-1}$, we regain
the Dirac equation, namely

$$\left(i\tilde{\gamma}^\mu \partial_\mu + \frac{m_0 c}{\hbar} \right) \tilde{\psi}_d(x) = 0.$$

where $\tilde{\psi}_d(x) = S\psi_d(x)$.

As the new set of gamma matrices are obtained from the original set
by a similarity transformation, therefore the anti-commutation relation
of Eq. (5.93) is preserved, namely

$$\{\tilde{\gamma}^\mu, \tilde{\gamma}^\nu\} = -2g^{\mu\nu}\mathbf{I}, \tag{5.101}$$

this enables us to show that each component in Dirac spinor $\tilde{\psi}_d$ also
satisfies the Klein-Gordon equation.

Let us return to the p-representation of the Dirac equation expressed
in Eq. (5.88), i.e.

$$\begin{pmatrix} 0 & \boldsymbol{P}_c \\ \boldsymbol{P} & 0 \end{pmatrix} \psi_d(x) - m_0 c \psi_d(x) = 0. \tag{5.102}$$

In the limit of zero mass, the equation reduces to:

$$\boldsymbol{P}_c \psi_l = 0, \quad \boldsymbol{P}\psi_r = 0. \tag{5.103}$$

Since in the zero mass limit, $p^0 = -p_0 = |\vec{p}|$ and the above equations can be expressed in terms of the helicity operator as follows:

$$(\boldsymbol{\sigma} \cdot \hat{\boldsymbol{p}})\psi_r = \psi_r, \quad (\boldsymbol{\sigma} \cdot \hat{\boldsymbol{p}})\psi_l = -\psi_l, \tag{5.104}$$

where $\boldsymbol{\sigma} \cdot \hat{\boldsymbol{p}} = \boldsymbol{\sigma} \cdot \boldsymbol{p}/|\boldsymbol{p}|$ is the **helicity operator**.

The equations imply that the left-handed spinor and the right-handed spinor are of opposite helicities.

5.6 Electromagnetic interaction and gyromagnetic ratio of the electron

Let us investigate a system of a spin $1/2$ charged particle moving in the electromagnetic field of the 4-vector potential $A^\mu(x) = (\varphi(x), \mathbf{A}(x))$. The principle of minimal interaction allows us to express the Dirac equation in the following form

$$\left\{ \gamma^\mu \left(P_\mu - \frac{q}{c} A_\mu(x) \right) - m_0 c \right\} \psi_d(x) = 0. \tag{5.105}$$

We shall adopt the Weyl representation of the gamma matrices and abbreviate $\boldsymbol{P} - q\boldsymbol{A}/c$ into the operator $\boldsymbol{\pi}$. Then the above equation can be separated into the following two coupled equations:

$$\left\{ \left(P_0 + \frac{q}{c}\varphi \right) - \boldsymbol{\sigma} \cdot \boldsymbol{\pi} \right\} \psi_l = m_0 c \psi_r, \tag{5.106a}$$

$$\left\{ \left(P_0 + \frac{q}{c}\varphi \right) + \boldsymbol{\sigma} \cdot \boldsymbol{\pi} \right\} \psi_r = m_0 c \psi_l. \tag{5.106b}$$

Let us apply the operator $\{(P_0 + q\varphi/c) - \boldsymbol{\sigma} \cdot \boldsymbol{\pi}\}$ upon both sides of the last equation and make use of the Eq.(5.106a). We then obtain the equation of the right-handed spinor as follows

$$\left\{ \left(P_0 + \frac{q\varphi}{c} \right) - \boldsymbol{\sigma} \cdot \boldsymbol{\pi} \right\} \left\{ \left(P_0 + \frac{q\varphi}{c} \right) + \boldsymbol{\sigma} \cdot \boldsymbol{\pi} \right\} \psi_r = (m_0 c)^2 \psi_r. \tag{5.107}$$

Consider for the case of pure magnetic field, namely $A^\mu = (0, \boldsymbol{A})$, then the stationary state solution of the last equation takes a much simpler expression, i.e.

$$(\boldsymbol{\sigma} \cdot \boldsymbol{\pi})^2 - \left\{ \left(\frac{E}{c}\right)^2 - (m_0 c)^2 \right\} \psi_r = 0, \qquad (5.108)$$

where we replace $P_0 = -P^0 = -E/c$. The first term $(\boldsymbol{\sigma} \cdot \boldsymbol{\pi})^2$ in Eq. (5.108) can be calculated as follows

$$(\boldsymbol{\sigma} \cdot \boldsymbol{\pi})^2 = \boldsymbol{\pi}^2 + i\boldsymbol{\sigma} \cdot \left(\boldsymbol{p} - \frac{q}{c}\boldsymbol{A}\right) \times \left(\boldsymbol{p} - \frac{q}{c}\boldsymbol{A}\right)$$

$$= \boldsymbol{\pi}^2 - \frac{q\hbar}{c}\boldsymbol{\sigma} \cdot (\nabla \times \boldsymbol{A}), \qquad (5.109)$$

and the second term $\left(\dfrac{E}{c}\right)^2 - (m_0 c)^2$ can be approximated as

$$\left(\frac{E}{c}\right)^2 - (m_0 c)^2 \simeq 2m_0 c \left(\frac{E}{c} - m_0 c\right) = 2m_0 E', \qquad (5.110)$$

in the non-relativistic limit, where E' represents the total energy minus the rest mass energy of the particle.

Finally Eq. (5.106a) reduces to the familiar expression of the Schrödinger equation for a spin $\frac{1}{2}$ charged particle coupled with the magnetic field $\boldsymbol{H}$, expressed as follows

$$\left\{ \frac{1}{2m_0} \left(\boldsymbol{P} - \frac{q}{c}\boldsymbol{A}\right)^2 - \frac{q\hbar}{2m_0 c}\boldsymbol{\sigma} \cdot \boldsymbol{H} \right\} \psi_r = E'\psi_r. \qquad (5.111)$$

The second term of the last equation stands for the energy of the intrinsic magnetic moment of the charged spin 1/2 particle $\boldsymbol{\mu}_e$ coupled to the magnetic field $\boldsymbol{H}$, namely

$$-\frac{q\hbar}{2m_0 c}\boldsymbol{\sigma} \cdot \boldsymbol{H} = -\frac{q}{m_0 c}\boldsymbol{S} \cdot \boldsymbol{H} = -\boldsymbol{\mu}_e \cdot \boldsymbol{H}. \qquad (5.112)$$

For the case of an electron, the **gyromagnetic ratio** γ_e, the ratio of the magnetic moment $\boldsymbol{\mu}_e$ to the **intrinsic spin** $\boldsymbol{S}$, is related to the g-factor in the following relation

$$\boldsymbol{\mu}_e = \gamma_e \boldsymbol{S} = -\frac{e}{2m_0 c} g_e \boldsymbol{S}. \tag{5.113}$$

Comparing Eq. (5.113) with Eq. (5.112), we reach a remarkable conclusion that, to the lowest order approximation, an electron has the g-factor equaling to 2, namely

$$g_e = 2. \tag{5.114}$$

The precision in the measurement of the electron magnetic moment had challenged theorists to higher order calculations in terms of fine structure constant, namely the radiative corrections, which are beyond the scope of our investigation.

5.7 Gamma matrix algebra and PCT in Dirac spinor system

Before entering into the discussion on discrete symmetry transformation of the Dirac spinor, we shall explore the properties of gamma matrices. As we have shown that the condition

$$\{\gamma^\mu, \gamma^\nu\} = -2g^{\mu\nu} \mathbf{I}$$

is valid for any representation of the Dirac matrix.

Out of these 4 Dirac matrices γ^μ, let us construct 16 matrices of 4×4 and divide them into 5 classes as follows:

(1) $\mathbf{I}$,

(2) γ^μ,

(3) $\gamma^5 = \gamma^0 \gamma^1 \gamma^2 \gamma^3$,

(4) $\gamma^5 \gamma^\mu$,

(5) $\sigma^{\mu\nu} = \frac{1}{2}[\gamma^m u, \gamma^\nu]$.

Class 1

It contains only the identity matrix, obtained, up to a sign, by making the square of any of γ^μ matrix. It commutes with all the gamma matrices in 5 classes.

Class 2

It contains 4 traceless matrices with the square equaling to $\pm\mathbf{I}$.

Class 3

It is a traceless, diagonal matrix obtained by successive multiplication of the 4 Dirac matrices. The matrix is denoted by γ^5 with the following properties,

$$(\gamma^5)^2 = -\mathbf{I},$$

$$\{\gamma^5, \gamma^\mu\} = 0.$$

Class 4

It contains 4 traceless matrices with the following properties,

$$\{\gamma^5\gamma^\mu, \gamma^5\gamma^\nu\} = 2g^{\mu\nu}\mathbf{I}.$$

Class 5

It contains 6 matrices, also traceless, by taking different values on μ and ν. The square of the matrix is, up to a sign, a unit matrix.

With all the preparation on the properties of the gamma matrices, we are now in position to discuss the parity transformation of the Dirac spinor. Firstly let us consider the parity transformation of a 4-vector R^μ in the Minkowski space expressed as follows

$$\mathcal{P}: \quad R^\mu = (R^0, \mathbf{R}) \xrightarrow{P.T} R^{p\mu} = (R^0, -\mathbf{R}),$$

$$\text{or} \quad \mathcal{P}: \quad \mathcal{P}R^\mu\mathcal{P}^{-1} = R^{p\mu}. \tag{5.115}$$

If R^μ is an operator such that $R^{p\mu} = (R^0, -\mathbf{R})$, then we call R^μ a 4-vector in Minkowski space.

In fact, there exists another vector S^μ called the axial 4-vector in Minkowski space if the transformation property takes the following relation

$$\mathcal{P}: \quad S^\mu = (R^0, \mathbf{S}) \xrightarrow{P.T.} S^{p\mu} = (S^0, \mathbf{S}),$$

$$\text{or} \quad \mathcal{P}: \quad \mathcal{P} S^\mu \mathcal{P}^{-1} = S^{p\mu} = (S^0, \mathbf{S}), \qquad (5.116)$$

when S^μ is an operator.

It is of the outmost interest to look into the parity transformation of a physical quantity S which is invariant under Lorentz transformation, namely a Lorentz scalar. We name S a scalar if it is also invariant under parity transformation, i.e.

$$\mathcal{P}: \quad S \xrightarrow{P.T.} S^p = S,$$

or if it is an operator, then

$$\mathcal{P}: \quad \mathcal{P} S \mathcal{P}^{-1} = S^p = S. \qquad (5.117)$$

In contrast to the last expression, if a physical quantity P is transformed by operator $\mathcal{P}$ as to meet the following relation, i.e.

$$\mathcal{P} P \mathcal{P}^{-1} = -P, \qquad (5.118)$$

then we call P a **pseudoscalar**.

Let us start to investigate the parity transformation property of the Dirac spinor. For the sake of simplicity, we shall express the Dirac equation in momentum operator as follows

$$(\gamma^\mu P_\mu - m_0 c)\psi_d = 0, \qquad (5.119a)$$

$$\text{or} \quad (\gamma^0 P_0 + \boldsymbol{\gamma} \cdot \mathbf{P} - m_0 c)\psi_d = 0. \qquad (5.119b)$$

Applying the parity operator $\mathcal{P}$ upon the Dirac equation above, and inserting the identity $\mathbf{I} = \mathcal{P}^{-1}\mathcal{P}$ right after the momentum operator P_μ, we obtain

$$(\gamma^\mu \mathbf{P} P_\mu \mathbf{P}^{-1} - m_0 c)\mathbf{P}\psi_d(x) = 0, \qquad (5.120a)$$

$$\text{or} \quad (\gamma^0 P_0 - \boldsymbol{\gamma} \cdot \boldsymbol{P} - m_0 c)\psi_d^{\boldsymbol{p}}(x) = 0, \qquad (5.120b)$$

where $\psi_d^{\boldsymbol{p}}(x) = \mathbf{P}\psi_d(x)$.

Therefore we have the following proposition:

Proposition 5.

Let $\psi_d(x)$ be the solution of the Dirac equation, and $\mathbf{P}\psi_d(x)$ be the parity transformed Dirac spinor, i.e. $\psi_d^{\boldsymbol{p}}(x) = \mathbf{P}\psi_d(x) = \psi(x^0, -\mathbf{x})$, then $\gamma^0\psi^{\boldsymbol{p}}(x)$ is the solution of the same Dirac equation.

The above conclusion is obvious if the Eq. (5.120b) is multiplied by γ^0 from the right, and making use of the relation $\gamma^0\boldsymbol{\gamma} = -\boldsymbol{\gamma}\gamma^0$, then we arrive at

$$(\gamma^\mu P_\mu - m_0 c)\gamma^0\psi_d^{\boldsymbol{p}}(x) = 0. \qquad (5.121)$$

Let us return to the 5 classes of the gamma matrices given at the beginning of this section. With the definition $\bar{\psi}(x) = \psi^\dagger(x)\gamma^0$, we shall investigate the parity transformation properties of the bilinear spinor combinations given as follows:

$$\bar{\psi}(x)\psi(x), \bar{\psi}(x)\gamma^\mu\psi(x), \bar{\psi}(x)\gamma^5\psi(x), \bar{\psi}(x)\gamma^5\gamma^\mu\psi(x), \bar{\psi}(x)\sigma^{\mu\nu}\psi(x).$$

To simplify the demonstration, let us take the solution to be non-degenerate, namely

$$\gamma^0\psi^{\boldsymbol{p}}(x) = e^{i\alpha}\psi(x).$$

Therefore one can easily verify that

$$\mathbf{P}\bar{\psi}(x)\psi(x)\mathbf{P}^{-1} = \bar{\psi}^{\boldsymbol{p}}(x)\psi^{\boldsymbol{p}}(x) = e^{-i\alpha}\psi^\dagger(x)\gamma^0 e^{i\alpha}\psi(x) = \bar{\psi}(x)\psi(x).$$

As for the transformation property of $\bar{\psi}(x)\gamma^5\psi(x)$, one finds that

$$\mathcal{P}\bar{\psi}(x)\gamma^5\psi(x)\mathcal{P}^{-1} = \bar{\psi}^{\mathcal{P}}(x)\gamma^5\psi^{\mathcal{P}}(x) = -\bar{\psi}(x)\gamma^5\psi(x).$$

It is because of these peculiar properties given above that we call $\bar{\psi}(x)\psi(x)$ a scalar and $\bar{\psi}(x)\gamma^5\psi(x)$ a pseudoscalar.

We shall summarize the results of the transformation properties for the rest of three bilinear spinor combinations as follows:

$$\mathcal{P}(\bar{\psi}(x)\gamma^\mu\psi(x))\mathcal{P}^{-1} = \bar{\psi}^{\mathcal{P}}(x)\gamma^\mu\psi^{\mathcal{P}}(x)$$

$$= (\bar{\psi}(x)\gamma^0\psi(x), -\bar{\psi}(x)\boldsymbol{\gamma}\psi(x)),$$

$$\mathcal{P}(\bar{\psi}(x)\gamma^5\gamma^\mu\psi(x))\mathcal{P}^{-1} = \bar{\psi}^{\mathcal{P}}(x)\gamma^5\gamma^\mu\psi^{\mathcal{P}}(x)$$

$$= (\bar{\psi}(x)\gamma^5\gamma^0\psi(x), +\bar{\psi}(x)\gamma^5\boldsymbol{\gamma}\psi(x)),$$

$$\mathcal{P}(\bar{\psi}(x)\sigma^{\mu\nu}\psi(x))\mathcal{P}^{-1} = \bar{\psi}^{\mathcal{P}}(x)\sigma^{\mu\nu}\psi^{\mathcal{P}}(x)$$

$$= (-\bar{\psi}(x)\sigma^{0i}\psi(x), +\bar{\psi}(x)\sigma^{ij}\psi(x)).$$

As one can also prove, it is of course much more complicated than in the case of parity that under the Lorentz transformation, the bilinear spinor combinations in class 1 or in class 3 behave as a Lorentz scalar, while class 2 or class 4 behave as a Lorentz vector and class 5 behave as a second rank antisymmetric tensor. We shall leave the proof to students of inquiring minds.

The charge conjugation property can only be analyzed in the system with the electromagnetic interaction. Let us express the Dirac equation for a particle with charge q as follows

$$\left\{ i\gamma^\mu \left(\partial_\mu - \frac{iq}{\hbar c}A_\mu \right) - m_0 c \right\} \psi(x) = 0. \qquad (5.122)$$

Denoting the charge conjugation operator by $\mathcal{C}$, and defining $\psi^c(x) = \mathcal{C}\psi(x)$, then the last equation can be cast into

$$\left\{ i\gamma^\mu \left(\partial_\mu + \frac{iq}{\hbar c} A_\mu \right) - m_0 c \right\} \psi^c(x) = 0, \qquad (5.123)$$

where $\mathcal{C}q\mathcal{C}^{-1} = -q$ is employed in obtaining the second term.

Taking the complex conjugate of Eq. (5.123), the equation becomes

$$\left\{ i\gamma^{\mu*} \left(\partial_\mu - \frac{iq}{\hbar c} A_\mu \right) + m_0 c \right\} \psi^{c*}(x) = 0,$$

which can be converted back into the Dirac equation

$$\left\{ i\gamma^\mu \left(\partial_\mu - \frac{iq}{\hbar c} A_\mu \right) - m_0 c \right\} \mathbf{C}\psi^{c*}(x) = 0,$$

where the solution is denoted by $\mathbf{C}\psi^{c*}(x)$ if the following condition is met, namely,

$$\mathbf{C}\gamma^{\mu*}\mathbf{C}^{-1} = -\gamma^\mu,$$

$$\mathbf{C}\psi^{c*}(x) = e^{i\beta}\psi(x),$$

with $e^{i\beta}$ standing for a phase factor and $\mathbf{C}$ a 4×4 matrix.

The solution of matrix $\mathbf{C}$ is strictly representation dependent. If we adopt γ^μ in Weyl representation, we have the solution $\mathbf{C} = \gamma^5\gamma^2$, because the Dirac matrices are of real matrix elements except γ^2. The charge conjugate spinor takes the following expression

$$\psi^C(x) = e^{-i\beta}\gamma^5\gamma^2\psi^*(x). \qquad (5.124)$$

As for the time reversal transformation of the following Dirac equation, one considers

$$\mathcal{T}\left\{\left(i\gamma^\mu\partial_\mu - \frac{m_0 c}{\hbar}\right)\right\}\mathcal{T}^{-1}\mathcal{T}\psi(x) = 0,$$

$$\text{or}\quad \left\{\left(-i\gamma^0\partial_0 + i\boldsymbol{\gamma}\cdot\boldsymbol{\partial}\right) - \frac{m_0 c}{\hbar}\right\}\psi^t(x) = 0, \tag{5.125}$$

where one takes the time t into $-t$ in the equation, $\mathcal{T}\psi(x) = \psi^t(x)$, i.e.

$$\mathcal{T}: \partial_0 \xoverset{T.R.}{\longmapsto} \partial_0' = -\partial_0.$$

Equation (5.125) can easily be converted back into the Dirac equation of the following expression

$$\left\{\left(i\gamma^0\partial_0 + i\boldsymbol{\gamma}\cdot\boldsymbol{\partial}\right) - \frac{m_0 c}{\hbar}\right\}\mathbf{T}\psi^t(x) = 0,$$

by multiplying a matrix $\mathbf{T}$, if the following conditions are met, i.e.

$$\mathbf{T}\gamma^0\mathbf{T}^{-1} = -\gamma^0,$$

$$\mathbf{T}\boldsymbol{\gamma}\mathbf{T}^{-1} = \boldsymbol{\gamma}.$$

We shall leave it to the readers to verify that $\mathbf{T} = \gamma^0\gamma^5$, and $\psi^t(x) = e^{-i\eta}\gamma^0\gamma^5\psi(x)$ with η being a phase factor.

5.8 Exercises

Ex 5.8.1

Show that a space-like vector is orthogonal to a time-like vector. (**Hint:** if $x^0 > |\vec{x}|$, then $y^0 < |\vec{y}|$ for $x \cdot y = 0$.)

Ex 5.8.2

Show that the sum of two time-like vectors in the same light cone is also a time-like vector.

(**Hint:** make use of the Schwarz inequality.)

Ex **5.8.3**

Show that two consecutive Lorentz boost transformations in the same direction with velocities β_1 and β_2 are equivalent to a boost with velocity β given by

$$\beta = \frac{\beta_1 + \beta_2}{1 + \beta_1\beta_2}.$$

Ex **5.8.4**

Verify that $D_{mm'}^{(\frac{1}{2},0)} = \gamma(m)D_{-m-m'}^{(0,\frac{1}{2})*}\gamma^*(m')$ can be cast in the matrix form as follows

$$D^{(\frac{1}{2},0)} = \begin{pmatrix} 0 & \gamma(\frac{1}{2}) \\ \gamma(-\frac{1}{2}) & 0 \end{pmatrix} D^{(0,\frac{1}{2})*} \begin{pmatrix} 0 & \gamma^*(-\frac{1}{2}) \\ \gamma^*(\frac{1}{2}) & 0 \end{pmatrix}.$$

Ex **5.8.5**

Verify that

$$\epsilon\sigma_i^*\epsilon^{-1} = \begin{pmatrix} 0 & 1 \\ -1 & 0 \end{pmatrix} \sigma_i^* \begin{pmatrix} 0 & -1 \\ 1 & 0 \end{pmatrix} = -\sigma_i.$$

Ex **5.8.6**

Let ψ and $\dot{\varphi}$ be the left-handed-spinor and co-right-handed spinor respectively. Prove then that $\dot{\varphi}\psi$ is an invariant scalar under Lorentz transformation.

Ex 5.8.7

Show that the Lorentz transformation matrix element Λ^μ_ν can be expressed as follows (**Hint:** by Eqs. (5.77) and (5.82).)

$$\Lambda^\mu_\nu = \frac{1}{2}\mathrm{Tr}\left\{\sigma^\mu_c D^{(\frac{1}{2},0)}\sigma_\nu D^{(\frac{1}{2},0)\dagger}\right\} = \frac{1}{2}\mathrm{Tr}\left\{\sigma^\mu D^{(0,\frac{1}{2})}\sigma_{c\nu} D^{(0,\frac{1}{2})\dagger}\right\}.$$

Ex 5.8.8

Show that $\boldsymbol{D}\gamma^\mu \boldsymbol{D}^{-1} = \Lambda^\mu_\nu \gamma^\nu$ where

$$\boldsymbol{D} = \begin{pmatrix} D^{(0,\frac{1}{2})} & 0 \\ 0 & D^{(\frac{1}{2},0)} \end{pmatrix} \quad \text{and} \quad \gamma^\nu = \begin{pmatrix} 0 & \sigma^\nu_c \\ \sigma^\nu & 0 \end{pmatrix}.$$

(**Hint:** since $\boldsymbol{X}' = \boldsymbol{D}\boldsymbol{X}\boldsymbol{D}^{-1}$ by definition.)

Ex 5.8.9

Define a new Dirac spinor as

$$\psi'_d(x) = e^{\frac{iq}{\hbar c}\alpha(x)}\psi_d(x).$$

Show that the Dirac equation for a charge particle with EM interaction is invariant under such gauge transformation. (**Hint:** with a new vector potential $A'_\mu(x) = A_\mu(x) - \partial_\mu\alpha(x)$.)

Bibliography

Dirac, P. A. M., *The Principles of Quantum Mechanics*, 4th Edition, (Oxford University Press, London, 1958).

Feynman, R. P. and A. P. Hibbs, *Quantum Mechanics and Path Integrals*, (McGraw-Hill, Inc. 1965).

Gürsey, Feza, Editor; *Group Theoretical Concepts and Methods in Elementary Particle Physics*, (Gordon and Breach, Science Publishers, Inc., New York, 1964).

Jordan, Thomas F., *Linear Operators for Quantum Mechanics*, 2nd Edition, (John Wiley & Sons, 1982).

Messiah, Albert, *Quantum Mechanics*, Vol. I & Vol. II, (North-Holland Publishing Co., Amsterdam; Interscience Publishers Inc., New York, 1962).

Riesz, F. and B. Sz-Nagy, *Functional Analysis*, translated by L. F. Boron, (F. Ungar Publishing Co., New York, 1955).

Tung, Wu-Ki, *Group Theory in Physics*, (World Scientific Publishing Co Pte Ltd., Singapore, 1985).

von Neumann, John, *Mathematical Foundations of Quantum Mechanics*, (Princeton University Press, Princeton, New Jersey, 1955).

Wigner, Eugene P., *Group Theory*, (Academic Press, New York, 1959).

Wybourne, Brian G., *Classical Groups for Physicists*, (John Wiley & Sons, 1974).

Index